AF769870

# WAR GAMES THROUGH THE AGES

*Books by Donald F. Featherstone*

**WAR GAMES**

**AIR WAR GAMES**

**NAVAL WAR GAMES**

**ADVANCED WAR GAMES**

**WAR GAME CAMPAIGNS**

# War Games through the Ages

## 3,000 BC to 1,500 AD

Donald F. Featherstone

Henry Regnery Company • Chicago

Copyright © 1972 by Donald F. Featherstone
Illustrations © 1972 by Stanley Paul & Co Ltd
All rights reserved.

First published in the United States by Henry Regnery Company,
114 West Illinois Street, Chicago, Illinois 60610
Published in Great Britain by Stanley Paul & Co Ltd
Printed in the United States of America
Library of Congress Catalog Card Number: 73-6448

# Contents

# War Games through the Ages

## 3,000 BC to 1,500 AD

# Introduction

A wargamer who takes up this hobby chooses a military period and then searches for information concerning the armies, their arms and equipment, dress and tactics of the period in question. If he selects a well documented period he will not have too much difficulty in obtaining basic information although no doubt he will spend many hours seeking the more technical or obscure details. Should he specialise in a less popular period, his research will take so much time that he will have little left to paint his armies and fight war games with them.

This book has been written to provide, under one cover, all the information that a wargamer requires to adequately build up armies in the Ancient and Medieval periods. Each period of warfare is considered and details given of the soldiers themselves, their dress, arms and equipment, leaders, tactics and their style of fighting. The major wars and battles together with the respective commanders are listed so that further information can be obtained by reading other accounts of these conflicts and the biographies of those leaders. Each section details the technique and styles of fighting of the various nations and armies and offers suggestions how this can best be simulated on a table-top battlefield.

This is very necessary because wargamers research their armies and paint them up, ensuring that every single facet and factor is as accurate as conceivably possible so that it represents, in miniature, a very reasonable simulation of what it actually was during that period of military history. But it all falls down at one critical point—unable to move under their own volition, the small model soldiers are strategically and tactically directed by the wargamer himself, who manoeuvres his armies with a military hindsight denied to their real-life commanders of long ago. This means that whilst his troops are accurate in every respect, they are performing tactical manoeuvres and evolutions far beyond the knowledge and comprehension of the men who commanded them in those far-off days. It is inconceivable to claim that any wargamer is capable of blotting from his mind ALL military knowledge of a later date than the period in which he is specialising. This means that all that is occurring on the table-top battlefield is a battle between men dressed as say Romans and Greeks, fighting a re-created or imaginative battle of their period but using tactics that would have done credit to Wellington in the Peninsula, Jackson in the Shenandoah Valley or Montgomery in Libya.

In many ways the Ancient and Medieval war games rules formulated by the War Games Research Group seek to remedy this situation by so formulating the conditions of battle that they reflect, to a certain extent, the peculiar attributes of the armies in question. That this is not a prime purpose of these rules would seem to be revealed by an introductory paragraph to the rules themselves:

These rules favour the general who acts offensively, seizes valuable terrain features, uses a wide-range and screen of light troops to cover his deployment and embarrass his opponents, who believes in closing with his enemy to obtain a decision, and who keeps a strong reserve under his hand to exploit opportunities and redeem failure. This calls for intelligent selection and combination of types of troops.

So, in a way, the wargamer is encouraged to select the army that most appeals to him and then use them in a manner best calculated to bring success but with little apparent bearing on their known real-life tactics. By specifying the standard tactical methods of fighting of every army of Ancient and Medieval times this book attempts to remedy that basic failing in the hobby of wargaming.

After studying the relevant section, the wargamer who favours a Roman army will know that they will advance in a certain formation, that they will hurl their pila and then come to close quarters with the gladus (short thrusting sword). After a simulated 15 minutes of fighting, they will be withdrawn and a fresh century or cohort thrown in to take their place. Their enemies, perhaps the Britons or Gauls, will fight in a desperate and ferocious fashion rather resembling the headlong charges of the Dervishes in the Sudan towards the end of the 19th century. Lacking discipline and order, with no tactical knowledge or control they did not attempt to outflank the enemy or to perform other such tactical manoeuvres. To fight with Romans, Britons or Gauls in any other manner is to prostitute military history so creating a completely false and unreal situaton. This book sets out the methods of fighting to which armies *should* conform although their wargamer/commanders are given a certain latitude that not only adds spice to the game but prevents it from taking on a stereotyped and forseeable aspect.

For each nation or army there is a Fighting Assessment Chart, listing twenty qualities that make an army a good, bad or indifferent force. By giving comparative points values to each of these qualities and then totalling them, a Fighting Assessment Figure is arrived at for each army. Thus, when handling a Viking force with a Fighting Assessment Figure of 45 against a Saxon force with a Fighting Assessment Figure of 23, a wargamer knows right from the start the comparative fighting abilities of each army and has a pretty fair idea of the degree of chance with which each goes into battle against the other. Not only is this Fighting Assessment Figure valuable in that respect but it also enables a singular facet of wargaming in the Ancient and Medieval periods to be employed with greater understanding and interest. The aspect under consideration, seemingly peculiar to wargaming in Ancient and Medieval periods, lies in the manner in which it is accepted that battles can take place between armies of widely diverging periods—the Persians and the Romans clash or the armies may contain a mixture of differing periods. Perhaps this is due to a mental telescoping of the centuries B.C. and A.D. and it is not something that seems to take place in other periods—no one takes on a Napoleonic force with a Marlburian army for example. By using Fighting Assessment Figures, it is made much easier to work out the comparative fighting values of armies who lived perhaps two or three centuries apart so that they can engage in table-top warfare without it appearing anomalous, incongruous or unworkable.

It is not easy to persuade a wargamer, however willing he may be at the start, to slavishly follow a course of action that is obviously doomed to failure. Perhaps there is an ironic lesson to be learned in this because that seems to be exactly what armies did do—it took the French nearly a hundred years to change the tactics that repeatedly brought disaster against the English longbow. But if a realistic war game is to be fought with armies performing in an accurate manner, then the wargamer must accept that what he gains on the swings he loses on the round-abouts. Whilst he is forced to use tactics that are obviously wrong to his modern military mind, the odds are balanced by him having a considerable numerical superiority, for example. Perhaps one of the strongest balancing factors is that of employing a system of 'grading' the commanders of the various armies so that, where they were known to have been exceptional men in real life (such as Hannibal or Julius Ceasar), then those commanders will be given a rating of 'Above-average'. Lesser commanders will be graded 'Average' or 'Below-average'. 'Above-average' commanders are given, within stated limits, the ability to undertake tactical movements perhaps outside the period—in other words, the wargamer is allowed to be himself. The 'Average' commander is given the ability, with certain reservations, to alter characteristic fighting patterns on limited occasions. The 'Below-average' commander is given no such opportunities and is compelled to fight his forces strictly within the known characteristic limits of their fighting tactics and ability. This is more fully detailed in the chapter 'Commanders'.

In spite of the hours that wargamers spend researching books and other sources of information in order to achieve accuracy in all aspects of their armies, it is easy to miss or overlook material perhaps written years before the wargamer was even born! There is much information available to the wargamer not contained in books but in the form of articles written by wargamers and historians and published in the various wargaming magazines and kindred publications. This book includes a bibliography for each historical section, listing the best known and most available books dealing with that period (including historical fiction) together with a listing of relevant articles or contributions that have appeared in military wargaming and historical magazines and journals over the past fifteen years. The majority of these magazines are well known and readily available to the wargamer, particularly if he is a member of a Club.

Details of the dress, arms and equipment of the various armies will be found in these books and articles and as a ready-to-hand easy reference source this book includes a list of references (giving illustration and page number, title of book, publisher, etc.) to coloured plates of the principal warriors of the armies of the Ancient and Middle Ages. In addition each chapter is embellished with a black and white line-drawing of a typical soldier of the period.

To solve the problem of availability of model soldiers for the Ancient and Medieval periods, an appendix lists those firms who can provide suitable figures. Every known manufacturer has been asked to supply this information and if there are any omissions it is simply because the firm concerned did not bother to answer!

Thanks are due to Paul Reeves who so ably carried out the photography for this book, as he has done for so many of my other books; to William Randell who has done the drawings and to Bill Jenkinson who is responsible for the diagrams. As a historical reference source, the book has undoubtedly benefited immensely from the advice offered by Bob O'Brien of the War

Games Research Group who 'vetted' the manuscript with considerable diligence. Neville Dickinson and Dick Higgs, of that most enterprising firm of model soldier manufacturers Miniature Figurines of Southampton, most willingly assembled and painted many of the various groups of figures for which I asked, to be photographed and decoratively included within the pages of this book. Thanks are also due to John Chapman and Robert Douglas who lent figures for this purpose.

D.F.

# Wargaming in the Ancient and Medieval Periods

The role of this book is to provide ready-to-hand information about Ancient and Medieval armies, their soldiers and their style of fighting so that suitable rules can be formulated or already established rules practised with greater understanding. Rules provide the conditions and limitations controlling the table-top manoeuvres of much scaled-down armies. More than that, if carefully and realistically compiled, they bring to the army concerned a singular personality and temperament of its own which with a little imagination can reflect many of the emotional and psychological factors that distinguished the armies and soldiers of the period.

Wargamers are fortunate that for the period covered by this book there is available a most comprehensive set of rules formulated by the War Games Research Group (composed of Philip Barker; Bob O'Brien; and Ed Smith) obtainable from Mr O'Brien at 75 Ardingly Drive, Goring-by-Sea, Sussex, England. These rules were formulated after discussion, testing, research and experiment covering a period of more than two years and have been amended in the light of experience gained since their publication. The compilers claim that '. . . great weight has been given to the characteristics of the different peoples who formed the warrior types of Ancient armies so as to bring out their varying fighting qualities and reactions to stress' obviously bring these rules well within the terms of reference of this book. See Appendix 6 for fuller details of these rules.

The Society of Ancients (Secretary, Philip Barker, 757 Pershore Road, Selly Park, Birmingham 29, England) are the most enthusiastic and well informed body of international wargamers specialising in the Ancient and Medieval periods. Membership of this Society should be a top priority for any wargamer interested in the period. The Society publish their own bi-monthly magazine *Slingshot* and can also supply very comprehensive sets of rules for wargaming in the period.

From the well-known international war games magazine *Wargamer's Newsletter* (69 Hill Lane, Southampton, SO1 5AD,

England) can be obtained booklets of rules covering the Ancient and Medieval periods. Although not so comprehensive or complex as the previous two sets, both give a realistic and reasonable game unlikely to frighten off the beginner.

In my book *War Games* (published by Stanley Paul) there is a complete section on Ancient warfare (pages 61–93). Here is given a set of reasonably simple rules followed by a demonstration of their use in a move-by-move account of the ancient battle of Trimsos—a purely fictional but highly enjoyable encounter in late B.C. or early A.D. period. As it involves both heavy and light infantry, heavy cavalry, mounted archers, chariots, war engines and elephants, the battle gives a fairly representative view of the majority of forces employed in Ancient warfare.

In that connection all the sets of rules mentioned detail conditions for the employment of these typically Ancient forces. Further details of the same nature may also be found in other books written by the author:

ELEPHANTS
*Wargames*, pages 71–3
*Battles with Model Soldiers* (David & Charles, 1970), pages 116–17, 124–5
Also see Dupuy, pages 40–1

CHARIOTS
*War Games*, pages 73–4
*Battles with Model Soldiers*, page 126

HORSE ARCHERS
*War Games*, page 65
*Battle with Model Soldiers*, page 133

MERCENARIES
*Advanced War Games* (Stanley Paul), chapter 21

No set of rules for Ancient wargaming is going to be realistic unless it considers the 'set' tactical formations common to many eras. In the War Games Research Group Rules, 'the number of ranks allotted to fight have been selected to encourage the use of tactical formations actually described in Ancient military manuals'. In addition to being described in the pages of this book, further information about them, in a wargaming sense, can be found in the following publications:

THE PHALANX
*Battles with Model Soldiers*, pages 115–16, 119, 123

ROMAN MANIPLE IN CHESS-BOARD FORMATION
*Battles with Model Soldiers*, page 118

ROMAN COHORT LEGION FORMATION
*Battles with Model Soldiers*, pages 120–1

SWISS PIKEMEN AND LANDSKNECHTE
*Battles with Model Soldiers*, page 134

Any of these formations simulated on a war games table will inevitably be so scaled down as to make them but a numerical fragment of their actual size. This fact, together with problems of a perhaps undulating terrain, make such formations extremely difficult to manoeuvre so that they bear even a coincidental resemblance to the formations they are intended to represent. It might not be possible for a force to adopt a formation from the very beginning of the battle so that fixing the troops on to already prepared movement trays beforehand is impracticable. The problem can be countered by cutting shapes to the required size from sticky-backed plastic tiles so that, when it is desired to place a force into a phalanx formation for example, the figures are quickly stuck on to the sticky surface of the tile which moves across the table complete with all its 'riders'. There are of course other movement tray methods involving sliding figures into slots prepared on movement cards, etc.

The majority of factors (or qualities) listed on the Fighting Assessment Chart are self-explanatory but there are those that can be amplified and advantageously built-in to rule books. For example, it can be ruled that the 'Desperation' factor is only invoked when the enemy inspire terror by being 'Ferocious'. A practical method of applying the 'Desperation' factor is explained on pages 105–6 of my book *War Games Campaigns* (published by Stanley Paul), 1970. In certain eras such as biblical times, early Greek and Japanese Samurai, Heroes or Champions went forward and challenged their opposite number to come out of the enemy ranks and meet them. This combat often had a decisive effect on the outcome of the battle but in any event it provides an interesting digression at some stage of a war game. The simplest way of deciding a combat between two figures on a war games table is to throw a dice, highest score winning. This system can be made more complex and realistic by the addition of 'Fighting Factors' similar to those described in the rules of the

War Games Research Group. A singular method which can be spread out to provide almost half an evening's wargaming entertainment is explained on pages 86–9 of my book *Advanced War Games*.

The factors of 'demoralisation' and 'panic' so closely resemble each as to make differentiation a little difficult. The author chooses the rule that demoralised units may not be rallied whilst panic-stricken units may be rallied—both methods of rallying to be performed within the rules and conditions under which the game is being fought. A number of wargamers have a ruling that an army panics when its casualties reduces its strength to 50% of its original size, then it runs from the field. A single unit can act similarly and other units in its vicinity have to decide whether or not they panic and follow suit.

Treachery has been with mankind since the beginnings of Time and while it may be controversial to assert that it was more common during the Ancient and Medieval periods, it would seem that this factor can be utilised to give a new facet to wargaming. The likelihood of treachery can be built in to a campaign narrative or, before a battle commences, some indication can be given that an army, part of an army or even an individual unit is suspected of treacherous motives and may well defect. In this case a commander may be given a 'Treachery Option' that he can exercise at any stage of the battle. To counter this some sort of a factor may be devised that enables the commander of an army to attempt to win back the loyalty of a treacherous unit. The effects of treachery could be that a treacherous unit or part of an army leaves the battle or it may defect to the enemy. In the event of a treacherous unit becoming panic-stricken so that it is routed, then it is beyond being rallied. As with most situations in wargaming, the easiest solution is to throw a dice but this relies far too much on luck to be the author's choice and the following 'Treachery Formula' is suggested.

First it has to be decided whether it is an entire army, a wing or a separate unit that is likely to display treachery. Having reached that conclusion then certain decisions have to be made:

(1) Is the commander of that force for or against treachery?

(2) Are the troops forming the force for or against treachery?

(3) Is the bribe offered to the unit high enough to persuade them to be treacherous?

These are basically yes-or-no choices and can be decided by the throw of a dice thus a dice scoring 1, 2 or 3 means that a commander is for treachery whilst a 4, 5 or 6 score means that he is against it. A score of 1, 2 or 3 means that his troops are for treachery, 4, 5 or 6 they are against it. A score of 1, 2 or 3 means that the bribe is high enough to make treachery attractive whilst a score of 4, 5 or 6 means that the bribe is not high enough.

Each 'yes' (for treachery) gives a −1 factor to the unit, each 'no' (against treachery) gives a +1 factor to the unit.

Now must be considered other relevant factors:

Does the unit have any national affinities to the people to whom it intends to defect? Yes −1; No +1.

Is its position on the battlefield conducive to defecting?

(a) It is on a flank and can get away −1
(b) It is flanked on both sides by other units and can only go forward +1
(c) It is not in a forward position and has other bodies of troops around it, making defection difficult +2.

(This situation might well be altered if any of the units around the one under consideration have already defected or indicated that they are likely to be treacherous.)

Are their side losing? Yes −1; No +1.

Having totalled up the pluses and minuses and arrived at a figure, throw two dice, one coloured and one white—take away the score on the coloured dice from the score on the white dice and add or subtract that figure to the total already arrived at. Then consult the following chart:

If resulting total after all factors have been considered and defensive dice added or subtracted equals 0 or a minus figure then the unit is treacherous and defects at charge-speed rate of movement and in any direction it wishes.

If the final score is plus one or more then the unit is not treacherous and carries on as before but if this figure only totals 1 or 2 then the unit will test for treachery again on the following move.

Accidental or unscheduled manoeuvres on the battlefield were, it must be confessed, a fairly frequent occurrence in Ancient and Medieval warfare. This fact has been accepted by the compilers of the War Games Research Group Rules so that their Reaction Effect Chart (pages 7–10) has facilities for uncontrolled advances. Other less complex methods of simulating this facet of warfare are the use of dice or Chance Cards. An unscheduled or accidental manoeuvre is a detrimental factor and cannot be used (within the framework of a set of wargaming rules) as a means of employing some tactical movement or action advantageous to the force making that movement.

Unscheduled and accidental manoeuvres can be controlled by strong leadership—perhaps it would never occur under an 'Above-average' commander.

Armies composed of different nations, tribes or factions each with their own commander under one supreme commander have never been particularly smooth working from the days of sticks and stones right through to Eisenhower and Montgomery in World War II. The potentialities for such discord were perhaps greater in Ancient and Medieval times than later although this is a matter of some controversy. Suffice to say that commanders of allied armies (particularly during the Feudal period) were prone to disagree about command and precedence, they disliked taking orders and conforming to plans not of their own devising. The handicaps suffered by the allies making up the Crusading armies when opposed to the unified command of the Saracens are a typical example. Rules should be formulated to allow for these authentic circumstances. They can be based upon the grading of the respective commanders or allied units could be placed in some numbered order of precedence and only allowed to move in that order whilst on the war games table. Factors can be inserted to cause groups to move outside their numbered order, so creating the accidental and unscheduled movements so typical of the period.

The Feudal period was a dark age so far as tactics were concerned (with the exception of the English during the Hundred Years War). Similarly, the Crusading knights took off into the blue at the first sight of the enemy just as did their descendants on a dozen battlefields of the Middle Ages. This can be reasonably simulated by:

(a) Feudal cavalry must always move at top speed, and without manoeuvring, towards the enemy.

(b) Feudal cavalry must always charge the nearest enemy unit within charge-move distance. A certain 'class-consciousness' would probably cause them to charge other cavalry rather than foot soldiers.

(c) If their own troops are in the way of the charge then the cavalry are highly likely to go through them and ride them down. It might be considered necessary to have a clause in the rules which saves the infantry from this ignominous fate.

To simulate the relatively 'wild' charges of cavalry almost throughout the whole of the period under review and at the same time to give opportunities for accidental or unscheduled manoeuvres, all cavalry units should be given a 'charge-radius' of 45° to either side of their formation or position at the start of the charge. An additional factor can be wedded to this by allowing commanders of 'Above-average' rating or, under certain conditions, commanders of 'Average' rating to allow a force to charge *as a whole* to one or other of the 45° angle of the 'charge-radius' (as opposed to the enforced 'straightforward' charge). This means that an entire force may start facing forward but may charge off in a body, either to right or left at an angle of up to 45°.

One of the simplest ways of simulating the tactical advantages due to a high-graded commander lies in the initial deployment of a force—in other words, one force (that of the inferior commander) is compelled to lay down first on the table so that the higher-graded commander can see their formation and make his plans accordingly.

It has been explained in the relevant sections of this book how, in these specific historical periods, battles were fought with more of an eye to material gain than to the actual destruction of the enemy. If the lower ranks of a defeated side knew that they were going to get their throats cut then with equal certainty their feudal lords were aware that, on payment of a calculated ransom, they would be returned to the bosom of their families until the next time they were captured. This business went a step further and, perhaps less magnaminously than it seems, even included the lower ranks when mercenaries and Condottiere took the field. The commander of a Free Company, having taken great pains to recruit skilled professional soldiers, had the most natural aversion towards losing them together with the fee per head he got for offering their services to whatever king or prince paid the most. And when mercenaries met mercenaries, both with the consciousness that today's enemies might be tomorrow's allies and the king you fought against today might be paying your wages next week, then the casualty rate was unnaturally low.

To translate this into wargaming terms, a commander in a strong position should be able to offer (within the framework of the rules) his less fortunate opponent the 'Honours of War' where the loser surrenders but retains his honour. One oft-voiced objection to wargaming lies in the fact that there are few if any provisions or indeed inducement to take prisoners—and this might well be the controlling factor in this case. A suggested method is to give each army a points total (one of the basic provisions of the Rules of the Ancient War Games Research Group). In this way the points value of a particular type of soldier when captured would be allotted to his captor and when the 'Honours of War' had been accepted then by negotiation a percentage of that total would be handed back to the surrendering side. For example, the War Games Research Group valuation for a Regular heavy cavalryman is 10 points—say ten such heavy cavalrymen have surrendered, thus giving a total of 100 points to their captors, but as they only surrendered on the condition that half of that total was returned to them then, by accepting the 'Honours of War', the captors have gained 50 points whilst the captives have similarly achieved 50 points. On the other hand, if a belligerent commander refuses the 'Honours of War' and his men are subsequently killed then the enemy gain not only the points value of each man killed but also a bonus of 50% of that total. Or it may be considered sufficient for the defeated side to lose the total points value whilst the triumphant army gain it.

From these suggestions arises the idea of a relatively new conception of wargaming—the manoeuvre game. Both armies are considered to consist entirely of mercenaries whose commanders are pledged to avoid killing the enemy's valuable soldiers and, in turn, to avoid losing their own highly priced commodity. Both agree to avail themselves of the 'Honours of War' system when the situation arises for its application. Under normal war games rules, each side tries to manoeuvre the enemy army or its separate units into untenable positions so that 'Honours of War' can be offered. The points value of each force are totalled at the start of the game and, on its conclusion, victory goes to the side with the highest points total.

One marked feature of Ancient warfare was the manner in which the losses of a defeated army were invariably so much greater than those of the victors. The massed ranks of Ancient armies were very vulnerable and when broken or assaulted from flank or rear it was rare that they could be rallied. Those who were able fled from the field while the others were either slaughtered or captured—few of the wounded survived in a defeated army. The victorious army usually had about five or six times as many wounded as killed. These factors should be reflected on the war games table where, all too often, a battle is allowed to end when one force is obviously going to win. To carry on to a logical and realistic conclusion, the victors should press home their assault and destroy their opponents—a particularly vital factor if the battle is part of a campaign.

Everything that is written in this chapter is put forward as a suggestion and a stimulus. Some of it has been conceived by the author, some of it has been adapted or inspired by the published ideas or comments of others. No apology is made for this because war games rules are largely an amalgam of those ideas from all sources that appeal to the compiler of a particular set of rules. This is really what wargaming is all about—unlike chess its rules are all things to all men and reflect the temperament, character and personality of their devisor, sometimes at the expense of realism and accuracy!

# Fighting Assessment

There are a number of inter-related factors which governed and controlled the fighting ability of the warriors of all the armies known to mankind and at every period of history. These factors outlined in the Fighting Assessment Chart probably apply to any period of military history but they are all certainly relevant when considering the soldiers of the Ancient and Medieval worlds. Every nation, every army and every force described within the pages of this book was controlled and affected to varying degrees by the qualities listed in our chart. Co-ordinated, the sum-total of all these qualities made him the good, bad or indifferent soldier that history records.

In an endeavour to reflect those qualities on paper to provide a workable system of matching in a table-top war game, one army or nation against another, the standard of each quality has been determined as follows:

| | |
|---|---|
| First Class | 3 points |
| Reasonable | 2 points |
| Little (Average) | 1 point |
| None | 0 point |

The last five factors (numbers 16–20) are considered to be detrimental qualities so the points values are reversed—that is to say a 'First Class' likelihood of demoralisation would score no points, whilst 'no' likelihood of demoralisation would score 3 points.

The detailed Fighting Assessment Chart is as follows:

*Fighting Assessment Chart*

| *Factor* | *Rating* | *Points* |
|---|---|---|
| 1. Ability to assault in controlled order | | |
| 2. Bravery | | |
| 3. Control | | |
| 4. Defensive ability | | |

| | |
|---|---|
| 5. Desperation | |
| 6. Discipline | |
| 7. Ferocity | |
| 8. Manoeuvrability | |
| 9. Mobility | |
| 10. Movement, speed of | |
| 11. Order | |
| 12. Tactical Sense | |
| 13. Training and Drilling | |
| 14. United action, capability for | |
| 15. Weapon Handling | |
| 16. Demoralisation, likelihood of | |
| 17. Impetuous Action, likelihood of | |
| 18. Panic, likelihood of | |
| 19. Reluctance to take definite action | |
| 20. Unscheduled Manoeuvres, likelihood of | |
| | *Fighting Assessment* |

Based on the belief that the Roman army, in its hey-day, was most likely to achieve almost 100% scoring in the majority of these factors, a Fighting Assessment Chart has been made out for each of the nations, armies or forces described in this book. Together with the 'Style of Fighting' of that force, this chart is set out in each section. This means that every nation or army has a Fighting Assessment Figure, serving as a means of grading that army and so indicating its chances against another force of the Ancient or Medieval world. For easy reference, the Fighting Assessment Figures of the nations considered in this book are as follows:

*Fighting Assessment Figure*

| | |
|---|---|
| Sumerians | 10 |
| Ancient Egyptians | 31 |
| Assyrians | 52 |
| Greeks | 32 |

| | | |
|---|---|---|
| Persians | 31 | |
| Macedonians | 44 | |
| Romans (all periods) | 55 | |
| Germans | 24 | |
| Carthaginians | 51 | |
| Gauls | 20 | |
| Goths | 30 | |
| Britons | 16 | |
| Barbarians (cavalry) | 39 | |
| Huns | 47 | |
| Mongols | 50 | |
| Saxons | 23 | |
| Vikings | 45 | |
| Normans | 28 | |
| Byzantines | 51 | |
| Chinese | 23 | |
| Franks | 32 | |
| Japanese | 19 | |
| Indians (Hindu) | 12 | |
| Crusaders | 30 | |
| Saracens | 39 | |
| Ottoman Turks | 35 | |
| Swiss | 54 | |
| Landsknechte | 50 | |
| Scots (14th century) | 23 | |
| French (14th/15th centuries) | Infantry: 7 | Cavalry: 23 |
| English (14th/15th centuries) | 48 | |

Some of the essential uses of this chart and its component factors are outlined elsewhere in this book and it is quite possible that other methods of use will suggest themselves to readers. One certain point arises—the basis for comparison between armies that it provides gives an almost fool-proof indication of the battle's outcome were they to be matched against each other with numerical equality and similarly graded commanders. For example, the Roman army with a Fighting Assessment Figure of 55 would massacre an army of its own size made up of Britons with a Fighting Assessment Figure of 16. In this case it can be seen that the fighting quality of the Roman soldier is about double that of the Briton, an historical fact which many systems of rules claiming authenticity must realistically simulate, as do the Ancient rules of the Ancient War Games Research Group.

The point at issue lies in the manner in which it is worked out—exactly how many more Britons are needed to form an army that can battle with a reasonable chance against a Roman force, so making for an enjoyable war game. In addition, such factors must be taken into consideration as balancing up those numbers with commanders of superior, average or inferior qualities. In this respect the Fighting Assessment Figure of each army may be used as a total and in comparison with other totals or else, by means of some formula, one Fighting Assessment Figure must be balanced against that of a possible opponent so as to give an 'odds' figure, i.e. how many Britons can one Roman legionary take on?

Each Fighting Assessment Figure must be put to practical use so that its claims are reflected in a realistic fashion. Perhaps the best and most simple way of doing this is to proportionally scale down a Fighting Assessment Figure so that, when added to weapon and tactical factors (see pages 13–16, Ancient War Games Research Group Rules), the resulting figure indicates the superiority or otherwise of the force concerned. To clarify, these weapon and fighting factors provide a series of plus or minus figures that reflect the power of the weapon concerned, the opponent and his weapon, plus such tactical factors as whether a unit is distracted or disorganised, whether it has received casualties itself or is firing upon a moving unit, etc. The resulting total, realistically effected by a dice throw to reflect the fluctuations of warfare, when applied to a Casualty Calculation Chart, gives the figure of enemy dead. If the figure, before being applied to the Casualty Calculation Chart, has added to it a further figure reflecting the fighting assessment value of the unit concerned, then the resulting casualties should realistically reflect the power of that unit. Similarly, a proportionate Fighting Assessment Figure added to a morale system such as that described in the Ancient War Games Research Group Rules or in that described in chapter 7 of my book *Advanced War Games* would similarly result in a realistic appreciation of the power of the unit concerned.

It is stressed that the Fighting Assessment Figure added to Fighting and Morale factors must be proportionate because in those cases it is single figures (and low ones at that) with which we are dealing. With the Fighting Assessment Figure of the Roman army, for example, being well over the fifty mark it will be necessary to give a representative figure such as 1 for every proportion of 200 points over and above 11 points. Thus, the Roman army with a Fighting Assessment Figure in the low 50s would have 1 point for the first 20 points, another point for the second 20 and a third point for the 12 or 13 remaining points—giving a total of 3 points to be added whenever weapon, firing, tactical or moral factors are being considered.

Elsewhere (in the chapter 'Commanders' for example) is mentioned the War Games Research Group method of writing down orders at the start of the game or the Featherstone method of writing them down at the start of each game-move. Fighting Assessment Figures can have a bearing upon either of these methods through armies with a high Fighting Assessment Figure being allowed to write orders each move and using an alternative order system; armies with an average Fighting Assessment Figure write their orders every third move and may use some alternative orders, whilst armies with a low Fighting Assessment Figure must write initial orders at the start of the battle for the entire game. This is, of course, similar to a suggested manner of working with graded commanders.

As it might not be entirely clear as to the reason for the inclusion of some of the factors or qualities that go to make up the Fighting Assessment Chart, there follows a briefly detailed explanation of each factor.

*Ability to assault in controlled order*

For a force to obediently move forward at the word of command or at the sound of a trumpet and, often in the face of missiles, advance steadily in an unbroken formation such as a phalanx or even in lines with each man keeping a set distance from his neighbour, not only requires great control, training and drill but also has a shattering morale effect upon the enemy. It has a big military bearing in separating the men from the boys and plays a vital part in the system of alternative orders and moves explained elsewhere in this book.

*Bravery*

Bravery is the ability to carry oneself defiantly and to encounter danger with courage—qualities inseparable from the good soldier. With such a vital bearing, its inclusion in this chart requires little further explanation.

*Control*

In a military sense, control means to dominate, to command, to direct or hold in check. These all are factors which repeatedly show up prominently in warfare of Ancient and Medieval times as they do in battles between disciplined troops and savages. That all these factors are inter-related is borne out by considering the often wasteful and fruitless bravery of undisciplined warriors when confronting the controlled bravery of well trained troops.

*Defensive ability*

This can be discussed in the terms of (1) a unit in a defensive role or (2) as the basic tactics of an army or force. It is in the latter sense that it should be considered here because a force or army noted for courageous if disorderly fighting methods (such as the Crusaders) displays poor defensive ability when under stress.

*Desperation*

It is not unreasonable to assume that, when faced with opponents noted for their ferocity, and with a stark realisation that torture and death are the only alternatives to victory, that a force will fight with desperation, resisting beyond its normal capabilities. A table-top simulation of this factor of desperation is explained in chapter 14, 'The Viking Raid', of my *War Games Campaigns*. Although desperation is usually invoked against 'ferocious' opponents, it could also apply when one or both sides realise that quarter is unlikely to be given. A typical example of such a situation could be the battle between Suetonius Paulinus' Roman army and Boadicea's Britons at Staines in 61 A.D. when the Romans were very conscious of the terrible things done by the Britons to Roman soldiers of other legions and the Britons knew that the Romans would exact the fiercest revenge.

*Discipline*

Discipline is mental and moral military training which brings soldiers under control and trains them to obedience and order. The discipline ingrained in a soldier through hours of drill and training is the greatest factor in making him stand when the going gets tough. It is also one of the most singular differences between trained and untrained warriors.

*Ferocity*

Throughout history certain warriors and nations have been held in the greatest dread by their opponents. In some cases their ferocity was a natural attribute whilst in others it was purposefully used as a military tactic. The first recorded people of this type were probably the Assyrians, others have been the Vikings (and their Berserks); the Huns and Mongols and, because they gave no quarter, the Swiss pikemen of the 15th century. Ferocity and the dread it inspires is a military factor that could have a battle half won before even the first blow had been struck.

*Manoeuvrability*

Generally considered to mean adroit and rapid movement of military formations, manoeuvrability can also be taken to represent the slow-moving, almost ponderous but excellently ordered and trained assault of the phalanx, the Scottish schiltron, the wedge of English longbowmen or the spear-point bristling formation of the Swiss pikemen. As opposed to the courageous but disorderly dash of tribesmen, the manoeuvre is an operation flowing fluidly in any direction and performed with order and control.

*Mobility*

Armies of Ancient times and the Middle Ages were not noted for the speed at which they moved around *en masse*. Consequently, those armies who displayed mobility were sufficiently few in number to be prominent. Among them were Attila's Huns and the Mongols of Genghis Khan and Tamerlaine who were reputedly vast in number, whereas it is now known that they were so mobile as to appear in one area when thought to be miles away, thus giving the impression of having numbers of armies and men scattered all over the place. Nomadic peoples, these cavalry armies, accompanied by their families and flocks, moved swiftly from place to place confounding and defeating their foes through their ease of movement coupled with their ferocity and fighting ability.

*Movement, speed of*

In 53 B.C. at Carrhae a Roman army commanded by Crassus was wiped out by the hit-and-run tactics of Parthian cavalry and later, in the Holy Land, the ponderous and slow-moving Crusaders were bewildered and dazzled by the swooping and wheeling of the Saracen horse archers. Speed of movement was a rare attribute in the wars of the Ancients and those nations who possessed light cavalry capable of carrying out these hit-and-run tactics chalked up a number of notable successes.

*Order*

When considering an army it is almost automatically thought of in good order or regular array, in a condition where every part or unit is in its right place. For obvious reasons, this state of affairs did not automatically apply with every army or force in the period we are now considering, and because order was more the exception than the rule, it is considered to be an asset of sufficient importance to merit a place in this chart. Military arrays capable of keeping good order were not always triumphant but they had a better chance than those in disorder.

*Tactical sense*

The ability to dispose military and naval forces in actual contact with the enemy so that they achieved some planned end was an art well known to many armies of the Ancient world but equally unknown to others. It is interesting to work through the armies set out in this book and note how tactical sense and ability is dotted about with chronological order. Armies were fighting to reasonably intelligent tactics in the years before Christ and yet (with the possible exception of Arsouf) the Crusaders displayed none whatsoever and even nearly 200 years later the French showed little tactical sense for the majority of the Hundred Years War.

*Training and drilling*

Whilst much of the art of war and handling of weapons came naturally to the warriors of the Ancient world such as the Barbarian cavalry hordes utilising their natural ability in the saddle, the most successful armies were those who had been trained and drilled. The Roman army is perhaps the best example of this and the hours of training spent by the English archer at the butts paid big dividends in actual warfare.

*United action, capability for*

In Ancient times and during the Middle Ages there were numerous instances of countries and tribes allying together for a common cause. Because of petty jealousies and pride, these alliances rarely worked and more often than not the commanders were trying to out-do each other rather than co-operate in bringing down the enemy. On a lesser scale can be seen centuries of contempt shown by cavalry for infantry and little attempt was made for co-operation or mutual support. On innumerable occasions even the most primitive plans were ruined by premature attacks or lack of movement by part of an army. Those nations and forces so bonded together by spirit backed with training and discipline making them capable of united action were usually the ones that came out on top.

*Weapon handling*

In the period under discussion it was natural for a man, particularly if he was of good birth or breeding, to be accustomed from birth to handling weapons. Even the peasant became adept with his spear and bow and arrow through the sheer necessity of hunting his food and it was not a very far step from stabbing or shooting a man as opposed to a deer. Roman proficiency with the pilum and the short stabbing sword, Saxon prowess with the battleaxe, English ability with the longbow and Swiss excellence with the pike made them masters of their craft and warriors to be feared.

*Demoralisation, likelihood of*

If the soldiers of an army are well trained and drilled, accustomed to discipline and control and are led by commanders with a reasonable tactical sense and ability in man-handling, then that army is unlikely to ever become demoralised. However, under certain drastic circumstances such as that undergone by the Romans at the hands of the Parthian horse archers at Carrhae or against the Germanic tribes in the Teutoburger Wald in 9 A.D., even troops so highly trained as the Romans had their discipline, cohesion, courage and endurance destroyed. Obviously, the less organised force, lacking discipline and training, will become demoralised quicker than the trained soldier or army.

*Impetuous action, likelihood of*

Perhaps the greatest examples of this are those demonstrated by the Crusading knights as they took off 'into the blue' against the elusive Saracen light cavalry. Similarly, the feudal knight, spurred on by his code of honour and his notions of chivalry, never failed to crash into action at the mere sight of the enemy. A typical example is that of the French knights at Crécy in 1346 when they actually rode down their own Genoese crossbowmen in their attempts to get at the English. Impetuosity, particularly when displayed by high-born and hard-to-control knights, presented commanders with a terrible problem because, like the boxer who loses his temper in the ring, the impetuous soldier leaves himself wide open.

*Panic, likelihood of*

Akin to demoralisation, panic is perhaps better able to be checked, and is also more likely to occur among ill-trained, poorly disciplined and badly led troops than among well controlled forces. A sudden infectious alarm leading to hasty measures, panic, causes unreasoning and excessive fright so that a force that has been fighting bravely will suddenly wilt and break with disastrous effects upon their comrades on either side of them. A typical example is that of the cavalry on the right wing of the Roman army at Adrianople in 378 A.D. wheeling away and fleeing from the field, leaving the legionaries to their fate. Most armies had a body of elite troops who never panicked and could be used to check that sad state of affairs in other units.

*Reluctance to take definite action*

As with so many other factors in this list, a reluctance to take definite action frequently stemmed from poor leadership, lack of control and order, and little tactical sense. On the other hand, it can obviously arise from a dread or fear of the opposition, to be particularly expected when the enemy are 'ferocious' warriors such as Vikings or Huns, etc. It would seem that the strongest factor in countering the reluctance to take definite action on the part of troops is good leadership but often this same reluctance arises in the mind of a commander particularly if he is in awe of the man who leads the enemy against him.

*Unscheduled manoeuvres, likelihood of*

An unscheduled manoeuvre is more a matter of timing than distance or direction. For example, a unit, part of a force or even the whole force may, particularly if coupled with a high 'impetuosity' factor, take off before their commander intends and at an unpropitious moment. On the other hand, if coupled with a high 'reluctance' factor, the force may delay its movement beyond the moment scheduled by their commander so that it goes at an equally unpropitious moment. Both may be effected by their commander's rating and his degree of control. An unscheduled manoeuvre may consist of a force deviating from its ordered line of attack, again this can be associated with other factors in this list.

*Note:* It must be borne in mind when considering Fighting Assessment Figures of the armies described in this book that they must, perforce, be of a rather nebulous nature. This is due to the fact that one Fighting Assessment Figure is being given to cover, in the majority of cases, the armies of a number of centuries. Of course, Fighting Assessment Charts can be made out for any specific army in any specific period but, in giving one figure to cover a nation's ability throughout its history the reader is stimulated and guided into deeper fields.

For example, the Chinese in some periods of their history were a power to be reckoned with and extended their sway well to the west of China, driving and breaking the nomads before them and keeping up this effort for about a hundred years. In the same way, the Mongols were only the force that they became when under the rule of someone such as Genghis Khan and his immediate successor. Without such leaders, while they still remained deadly fighters, they tended to dissipate their strength in ceaseless feuding and raiding amongst their own kind and were not averse to joining forces with an invader if they thought that by so doing they could gain the advantage over their local enemies. Therefore, in such periods their Fighting Assessment would be reduced accordingly as their capacity for United Action, Order and Discipline would become low with tribe fighting tribe and clan against clan. Similar arguments can be advanced for most of the other types and races.

In working out Fighting Assessment Figures given in this book, an attempt has been made to assess trends, record and ability over the entire period in question. At the same time, there must inevitably be the temptation (which is confessed has been freely succumbed to) to judge the army as it was at its peak. This is not unreasonable in two senses: (1) because it is assumed that the majority of wargamers want that army when it was *good* and (2) because the 'opposition' must also be presumed to be at its best unless it is intended to simulate the actual period of history when one army was at its peak whilst its historical opponents were going through a bad time.

Some of the Fighting Assessment Figures may be considered with a sceptical eye in the sense that, when compared with other nations at distances of even centuries, they would appear to have been equally effective fighting forces. It must be borne in mind that, when assessing the fighting value of an army, that figure must inevitably be coloured or affected by its known opposition.

Perforce, these Fighting Assessment Charts are worked out on the basis of a 'neutral' terrain. It has been suggested that it would be interesting to see how they could be amended in many instances when one side was operating in terrain more suitable to itself, with perhaps the advantage of a friendly local population. Whilst these are considerations that would apply mainly to campaign games, they could be brought into the normal war game encounter by setting up situations and working out the outcome of the war games table. This would be a popular and welcome change from the straightforward confrontation that so often ensues. A supreme example is the destruction of Varus's legions in Germany where the German forces would have little chance of a straightforward battle with the legions on anything like open ground. This question of favourable terrain could be almost decisive where cavalry armies are concerned as even the best of them lost a major part of their effectiveness when they tried to operate on unsuitable country whereas infantry-based armies could be equally unhappy in this period when opposing them in open country.

# Commanders

It is very difficult to attempt, as this book suggests, to restrict a wargamer to manoeuvring his forces in a manner that is typical and strictly accurate for their period and type. With the best will in the world that wargamer is likely to be adverse to throwing, willy-nilly, his beautifully painted ancient Britons at the impregnable Roman formation waiting to receive them—and yet that was the way it went in those far off days!

The system which probably holds the most promise in enforcing these conditions is that of graded commanders allowing 'Above-average' commanders, 'Average' commanders and 'Below-average' commanders (in the person of the actual wargamer himself) to control their forces in a specified fashion. In other words, the 'Below-average' commander can only control and manoeuvre his forces *strictly* in the terms of their 'Style of Fighting' without any deviation or allowances whatsoever.

The 'Average' commander similarly has to control and manoeuvre his force along the lines laid down in their 'Style of Fighting' but with certain allowances.

The 'Above-average' commander represents such outstanding leaders as Alexander, Hannibal, Julius Caesar and Belisarius, etc. These leaders are represented by the wargamer himself, who is able to control and manoeuvre his forces, without any period-restriction, taking full benefit of the hindsight of military knowledge gained by reading books of tactics devised centuries after these leaders had become dust. He is also given certain other allowances.

If the 'Below-average' commander is going to labour under such disadvantages, then what inducement is there for the wargamer to act as one of them? There are a number of answers to this problem; for example in a case where an absolute novice wargamer is fighting against a highly experienced practitioner, then the novice might well be considered to be a 'Below-average' commander whilst the experienced player would at least be an 'Average' if not an 'Above-average' commander. This has built-in difficulties arising from the fact that a 'Below-average' commander must necessarily have a pretty reasonable knowledge of wargaming to be able to make his forces conform to their rigidly laid-down 'Style of Fighting' but this is not a problem that defies solution. Then there is the situation in which history itself dictates that one commander shall be 'Below-average' whilst the other shall be 'Above-average'—such a situation occurred on many occasions when Julius Caesar was invading Gaul or Britain whilst a classic case may well be when Alexander defeated Porus at the Battle of the Jhelum in 326 B.C. On the other hand, a war game may be arranged with no historical connections other than that the armies are garbed and equipped and fight in a style of long ago. Then it may well be that the wargamers involved wish to add spice to the game by pitting a small force under an 'Above-average' commander against a much larger force led by a 'Below-average' commander—how then does one balance those forces so as to give a reasonable battle? On page 59 of my book *War Games*, a method is described giving points values to various types of troops (in this case infantry=1 point, cavalry 2 points and a gun 10 points whilst an army consisted of 250 points). Taken further, this method can be used to determine the rating of one's commander. This method has been brought to a most satisfactory state for our purpose in the 'War Games Rules 1000 B.C. to 500 A.D.', compiled by the Ancient War Games Research Group. Their chart of relative points values (on page 3 of their book of rules) allocates points values to various classifications of troops, differentiating between Regulars and Barbarians. Suffice to say for our purpose, the chart gives the Commander-in-Chief a value of 100 points (for the purposes of comparison, a Roman legionary in leather armour is given a value of 4 points) so, by giving 'Above-average' generals a higher points value and 'Below-average' generals a low points value, the deficiencies of the latter are in a realistic fashion partially balanced by the extra troops that can be 'bought' with the surplus points.

The War Games Research Group rules, differentiating between the various grades of commander, give these commanders a definite role to play. He is able to give orders *direct* to any unit to which he has attached himself, as opposed to the pre-written order system embodied in these rules and described elsewhere. By being individually with the unit, the commander gives that unit an increased standard of morale and generally adds to its steadiness. Obviously, these factors will vary in cases where generals are considered to be 'Above-average', 'Average' or 'Below-average'.

The rules of the Ancient War Games Research Group provide

for the wargamer writing down, at the beginning of the game, the orders that each unit will normally follow and to which they must conform until fresh orders are received by means of messengers, direct contact by the commander himself or by pre-arranged visual signals. Units follow these orders throughout the game until they are changed by a fresh order or until they are changed by some enforced action that has occurred or by the result of a Reaction Test. Experience shows that it is rarely easy to send fresh orders to units when using this system of rules—a fact which leads to one method of differentiating between various grades of commander.

It has already been stated that 'Below-average' commanders must conform rigidly to the 'Style of Fighting' of the force they command so that, just as in the Ancient War Games Research Group rules, these orders remain in being until altered by some or other of the factors already mentioned. 'Average' commanders could be given the benefit of being permitted to write orders for each unit to take them through, say, three moves of the game. At the conclusion of the third move, fresh orders may be written if changes of orders or circumstances have not already altered the original instructions. 'Above-average' commanders may be permitted to write orders at the beginning of *each* game-move in the manner described on pages 19–21 of my book *Advanced War Games*.

Another derivation would be to allow commanders, according to their rating, to write down orders at the beginning of the game but to change those orders if they desire, taking a certain number of moves to do so. In other words, the lower the commander's grade, the greater number of moves he takes to alter his orders and, conversely, the 'Above-average' commander can change his orders much more rapidly. Taking this further, the system of alternative moves and orders offers itself as a really tangible way in which to reasonably emphasise the difference in gradings between commanders. 'Below-average' commanders are *never* allowed to avail themselves of alternative moves and orders; 'Average' commanders may use one (or any other pre-determined and agreed number) alternative moves or orders whilst 'Above-average' commanders may use as many as they desire.

# Alternative Moves and Orders

A certain proportion of the move is taken up by the commander (or the force) assessing a new situation and responding to it.

This proportion depends upon:

(a) The commander's rating.

(b) The Fighting Assessment Figure of the force.

(c) The Fighting Assessment Factor involved, i.e. a force is going to charge but are themselves charged—if their 'ability to assault in controlled order' factor is high then they are considered to be sufficiently well organised to gather themselves together and carry on with their charge without delay. If the factor is low, then they waver, so taking up a proportion of their move, possibly causing them to be hit under disadvantageous circumstances. Should the force, seeing they are going to be charged, wish to take the charge in a defensive formation then two factors are involved:

(1) Capability for United Action

(2) Defensive ability

If (1) is high then the force, without delay, assumes a defensive formation. If it is low, then it is indecisive, so taking up part of the move and perhaps causing it to be hit whilst in disorder.

In (2) see section dealing with Defensive Ability (page 29).

In any event, these features have to be considered by BOTH sides and are possibly susceptible to commander's rating and also Chance Cards.

*Methods of employing alternative moves and orders*

(1) When moves are written down for the next move at the end of each preceding move (described on pages 19–21 of *Advanced War Games*), the wargamer either writes his intended move and then marks it with 'X' or 'A/O' (Alternative Order) writing that move on a separate slip of paper or else he writes *both* moves in the space provided on the movement chart, clearly showing that they are separate by dividing them with a line, etc. Considering the limited space available, this could cause confusion, so a code can be used—suggested alternative moves are listed below, together with code-names.

(i) Original order is to move forward 12 inches but it is desired to guard against an attack coming in from a flank. Thus:

| 12″ A/O F.I.A. |
| --- |

Meaning 'Flank if attacked' F.I.A. allows the unit to re-form (during its move) facing to the required flank from which the attack is coming.

(ii) Original order requires the unit to remain stationary but the enemy charges it. If this charge were received in a stationary position its effect would be increased (particularly should either force be cavalry). So the orders are written as follows:

| Stay A/O C.C. |
| --- |

This means that the unit, seeing its attackers, may counter-charge them. An interesting development here would arise if the charge is coming in from a flank, because the A/O could be F.I.A. but the charge would then have to be taken at the halt! In other words, an A/O does not countermand a charge from a flank, but only an attack delivered within the normal 45° 'charge-radius' of the attack unit. A successful A/O 'C.C.' means that *both* units receive the usual bonus for impetus or shock (detailed on pages 66–7 *War Games* and page 89 *Advanced War Games*).

(iii) As before, a unit is unexpectedly charged, but on this occasion its original orders were to 'move forward 6 inches.' The necessary A/O would be:

| 6″ A/O C.C. |
| --- |

There is another manner of coping with a charge which gains the advantages of firing upon the charging enemy but, in the event of that fire *not* turning-back the charge, the enemy have to be received by a unit that is only beginning to move and not actually counter-charging. This means that the charging unit gains a 'shock' bonus whilst the unit receiving the charge does not.

In this case, the A/O. is thus:

| Stay A/O ¼ M.I.M. |
| --- |

In other words, the original order was to stay put (or it could be a move) with a proviso that, if attacked, the defending unit may stay put for $\frac{3}{4}$ (or $\frac{1}{2}$ if desired) of their move and then move forward, so as not to be caught 'flat-footed' for the remaining $\frac{1}{4}$ (or $\frac{1}{2}$) move. The benefit lies in the fact that the defenders, if they are missile-men, may fire upon their attackers in the hope of turning them back or reducing their strength for the forthcoming mêlée. At the same time, any other missile men not actually being attacked may, if within range, fire upon the attackers without their fire being masked by the defenders advancing early in the move.

(iv) This A/O is slightly different in that it permits a force to be stationary in reserve and yet be able to go to the assistance of a friendly unit that is suddenly attacked. It is written:

| | | |
|---|---|---|
| Stay | A/O | S meaning 'Support'. |

Thus, if any unit is attacked the supporting unit may go to its assistance *providing* the supporting unit is facing in the required direction or can spare enough of the move to take $\frac{1}{4}$ of it to turn in that direction. This latter course is not often possible because the supporting unit can only begin its life-saving move when the attackers have advanced $\frac{1}{4}$ of their charge-distance.

(v) Similarly, an A/O can be given to attackers so that they, deciding discretion to be the better part of valour, can change their minds about charging if they see large numbers of supporting troops massing against them, for example, thus:

| | | |
|---|---|---|
| Charge<br>'X' Unit | A/O | W.I.D. |

This means that the charging-unit may 'withdraw if desired' after it has done $\frac{1}{4}$ of its charge-distance *but* it loses this option once this attacking force has covered more than $\frac{1}{2}$ of its distance. To withdraw, the unit is wheeled (if under good control), or pulled up and turned, which takes 3 inches of the move. Forces subject to 'Feudal' attributes are never allowed to halt a charge in this manner.

These are just a few samples of Alternative Orders—many other versions can be devised through experience and to suit specific situations. To avoid acrimony, it is strongly advised that ALL participants in a particular war game are aware of every alternative order and its implications.

Research Group Rules incorporate optional actions proper to certain types in certain situations, e.g. cavalry may countercharge attacking enemy cavalry, subject to proper reaction and ability to wheel in time. Their orders may of course specifically forbid it.

It cannot be too strongly emphasised that everything suggested in this section is subject to possible restrictions of a particular force's Fighting Assessment Figure besides having to fall within the framework of that force's 'Style of Fighting'.

There are other lines of thought which may be followed in this matter of graded commanders. For example, a commander's rating may, if desired, add or subtract from the 'Fighting Assessment Figure' of a force or effect the individual points ratings for Fighting Assessment Factors. This is something that must be worked out by the individual wargamer because it goes against the precept already laid down in this book of having a Fighting Assessment Figure for each type and period of warrior.

A less complicated method of differentiating between the grades of commanders can be to allow 'Above-average' commanders to be represented by the wargamer himself and to act as he thinks fit whilst 'Below-average' commanders must rigidly conform to the laid-down style of fighting of the force he controls. But the 'Average' commander is controlled by a method that gives him a reasonable scope for becoming, as might well be the case, an 'Average' commander with flashes of brilliance or an 'Average' commander who has his 'bad' days. The method suggested is to make up a series of Chance Cards (for details of the use of Chance Cards read chapter 21, 'The Shenandoah Valley Campaign' in my *Campaigns with Model Soldiers* and pages 28, 137–8 of *Advanced War Games*). In this case it is suggested that the cards are made in a series to conform to the twenty factors shown on the Fighting Assessment Chart and be so worded as to give an advantage or a disadvantage. The 'Average' commander would pick a card when a situation arises that comes under a specific Fighting Assessment heading, i.e. when a force with a low rating of the 'ability to assault in controlled order' finds it necessary to make an attack. The card may well 'up' the ability of that unit to assault in controlled order with advantageous effect. On the other hand, the card may reduce the unit's ability to do this. A further refinement can allow for any Fighting Assessment Factor of 2 or 3 to allow the card to be disregarded if the 'Average' commander desires. For example, a unit with a reasonable ability to assault in controlled order (2 points) might draw a Chance Card in the hope of 'upping' that ability to first class (3 points). INSTEAD, the card decreases the unit's ability to assault in controlled order from 'reasonable' down to 'little' ability but as the unit's initial rating was 2, the 'Average' commander is able to disregard this disadvantageous card.

For general details on the grading of commanders, see chapter 18, 'Commanders Good and Bad', of my book *Advanced War Games*.

Medieval amendments to War Games Research Group rules allow for higher value command figures, with varying effects on fighting and morale. They have to be 'bought' for higher points values.

# THE SUMERIANS
## *The Beginnings of War*

The club and spear of Primitive Man were primarily used for hunting—but they could kill an enemy. In the beginning, local brawls and scuffles were spontaneous interruptions of a peaceful existence, but as Neolithic Man progressed, villages became towns and towns grew into cities, leading to disputes over boundaries, property and trade-routes. Men learned to work metals and in the Bronze Age came the first offensive weapons—the dagger, lengthened to a sword and attached to a pole as a spear.

It was in Sumer (the lower part of Mesopotamia) that fighting turned from a sporadic into a systematic activity. Civilisation first arose in this area in about 3000 B.C.; later, the Sumerian cities began to fight each other as war leaders led forth their citizen-armies. Consolidating their power, these leaders established a hereditary kingship which required the backing of a military establishment. And so the institution of war was bestowed upon mankind.

The Sumerian foot soldiers of about 2900 B.C. wore leather helmets, carried heavy shields, were armed with bow and spear and marched in ranks six-deep. The sword, the bow, the axe, the dagger and the spear were the principal weapons of warfare for thousands of years. The ancient *Epic of Gilgamesh* tells how they were first used in Sumeria:

> And now they brought to them the weapons, they put in their hands the great swords in their golden scabbards, and the bow and the quiver. Gilgamesh took the axe, he slung the quiver from his shoulder, and the bow of Anshan, and buckled the sword to his belt; and so they were armed and ready for the journey.

Paintings on pottery tell us that the chariot was used by the Sumerians, probably more as a means of transport to carry a chief or champion into battle than as an actual weapon. Drawn by onagers (a species of ass) they had solid wheels and were clumsy and slow. Moving on foot, the fighting man armed himself as best he could and followed the chieftain into battle as part of a disorderly mob. With the passing of the centuries, the disadvantages of these motley gatherings were countered by separating the men armed with various weapons into groups so that spearmen and axemen fought alongside comrades similarly armed whilst bowmen went forward in groups perhaps as skirmishers in front of the main line. This encouraged a more 'professional' approach so that spearmen acquired helmets and shields and, later, body armour of bronze, leather, quilted cotton and eventually iron.

These unorganised and undisciplined levies were almost impossible to control and it is unlikely that they moved and fought in any rigid formation. Urged and pushed into position by their leaders, they straggled and jostled into a ragged battleline with the more ostentatiously bold warriors loudly demanding a place in the front.

In time, their leaders became less tactical commanders than champions, going forward as did Goliath to encounter an enemy leader in mortal combat, surrounded by their followers. The outcome of these single combats often decided the day in their particular part of the field because when one specific leader fell his followers were most likely to rush back to safety, often bearing the body of their leader. Sometimes a leader, if finding himself hard pressed or out-matched, would flee the field by jumping into a conveniently placed chariot and hastily leaving the scene of action. On such occasions it is likely that his followers sullenly retreated with him, after growling defiance and belligerence at the triumphant enemy facing them.

United after centuries of inter-city strife, the Sumerians were conquered in about 2750 B.C. by the Akkadian, Sargon the First. The United Empire that resulted was soon conquered by the Amorites who founded Babylon on the Euphrates. Under Hannurabi the First, by 2100 B.C. they had conquered most of the land between the rivers.

### STYLE OF FIGHTING

A disorderly, disorganised, undisciplined mob of levies, including bowmen, with little organisation. Their sole objective was to overwhelm the enemy before he managed to move superior numbers of men so as to prevail. With the passing of time, a modicum of organisations separated them into groups of spearmen, axemen, archers, etc. Individual champions were in vogue.

*Fighting Assessment Chart:* SUMERIANS

| | *Factor* | *Rating* | *Points* |
|---|---|---|---|
| 1. | Ability to assault in controlled order | None | 0 |
| 2. | Bravery | Reasonable | 2 |
| 3. | Control | None | 0 |
| 4. | Defensive ability | None | 0 |
| 5. | Desperation | Not applicable | — |
| 6. | Discipline | Little | 1 |
| 7. | Ferocity | None | 0 |
| 8. | Manoeuvrability | Little | 1 |
| 9. | Mobility | Little | 1 |
| 10. | Movement, speed of | Little | 1 |
| 11. | Order | Little | 1 |
| 12. | Tactical Sense | None | 0 |
| 13. | Training and Drilling | None | 0 |
| 14. | United action, capability for | None | 0 |
| 15. | Weapon Handling | Reasonable | 2 |
| 16. | Demoralisation, likelihood of | First Class | 0 |
| 17. | Impetuous Action, likelihood of | First Class | 0 |
| 18. | Panic, likelihood of | First Class | 0 |
| 19. | Reluctance to take definite action | Likely | 1 |
| 20. | Unscheduled Manoeuvres, likelihood of | First Class | 0 |
| | | *Fighting Assessment* | 10 |

# THE ANCIENT EGYPTIANS

Occupying the geographical location that was favourable to the peaceful development of their civilisation, the Egyptians built their pyramids disturbed only by small-scale military activities such as those between border guards and raiding desert nomads. In about 1750 B.C. Egypt was invaded by a barbarous people, the Hyksos. Known as the 'Shepherd Kings', their origin is in doubt and they are believed by some to be Israelites or to be akin to the Hittites. The Hyksos fought from chariots manned by fast-firing archers. In the process of driving them out, Egypt became militaristic for the first time, organising a national army in two main divisions, north and south. Realising that chariots changed the tactics of war by making battles mobile and fast moving, the Egyptians equipped their forces with war chariots made entirely of wood and leather; light vehicles with spoked wheels capable of manoeuvring at high speed and so light that a man could carry one on his shoulders.

The first great Egyptian General was the Pharaoh Thothmes III who ruled for 50 years from about 1500 B.C. He greatly expanded the small Egyptian standing army, grouped infantry into units with standards each in the form of some sacred object or device borne upon the points of a spear by an officer chosen for his bravery. Mercenaries were employed and units using different weapons were trained to co-operate with each other. There were units of archers and slingers, heavy infantry and spearmen who carried large shields that covered them from neck to ankle. Little is known of the use of cavalry by the Egyptians. With these forces, Thothmes III fought the battle of Megiddo, better known as the Biblical Armageddon. It is said that this well disciplined Egyptian army was the first fighting force trained to respond to the sound of trumpet or drum.

It was under another Pharaoh, Ramses II, that the army of ancient Egypt reached its peak. It is recorded that it carried out the first known flanking movement at the battle of Kadesh in 1286 B.C. Chariots played a great part in Egyptian warfare of this period; used in large numbers, they relied on shock effect as well as fire-power from the archer carried in each vehicle. It was a time when, against any but highly disciplined infantry, the massed chariot attacks were most effective, only being checked by counter-charges which resulted in a mêlée of rushing vehicles flashing past each other in a cloud of dust, arrows and javelins.

Military uniforms were simple in style, consisting of a striped bonnet head-dress and a waist girdle of strips of leather sewn to one another. Iron or bronzed helmets and leather jackets covered with metal scales were reserved for higher ranks and the mercenaries. The ordinary soldiers carried an ox-hide shield stretched on a light wooden frame with a metal disc about eight inches across in the middle and endeavoured to take the impact of arrows and spears upon this metal disc. In addition to the khopshou, a hatchet with a broad curved blade, the soldiers were variously equipped with axes, bows, daggers, short swords, javelins and the six-foot-long short spear. On the battlefield, light infantry armed with bows slings and daggers went forward in advance endeavouring to disorder the enemy with a hail of missiles. Next the heavy infantry, their spears held out in front of them, drove like a wedge into the enemy's centre. At the same time, massed chariots on either flank bore down at full gallop in a charge so perfectly timed and drilled that not a single chariot would break the line.

All the armies of the Pharaohs had their specialist bodies of pioneers and sappers, who built the camps which served the armies for the duration of the campaign. Consisting of a big rectangle surrounded by a deep wide ditch and a wall of earth protected by shields, the camp was approached by a bridge leading to a single door guarded by sentries. The Egyptians carried out siege operations, using their archers to keep the enemy from manning the walls whilst the assault troops got to close quarters—the Sudanese Negroes were the best troops for this purpose. The Sappers undermined the walls, protected from the enemy missiles by wooden huts covered with hides, and scaling ladders were mounted so that sieges were prosecuted in a manner that was to become traditional.

The Egyptians fought their greatest campaigns on Syrian territory against the Hittites, a warlike mountain people of Indo-European origin who briefly created an empire in Asia Minor and Syria. Using tactics that mastered the horse-drawn chariot, at the battle of Kadesh the Hittites led Ramses II into a trap from which he was barely able to extricate himself. Although Ramses went home boasting of victory, Egypt's power waned from this date onwards.

There was an increasing use of foreign mercenaries (i.e. Greeks) and other 'outlanders' from 1200 B.C. onwards.

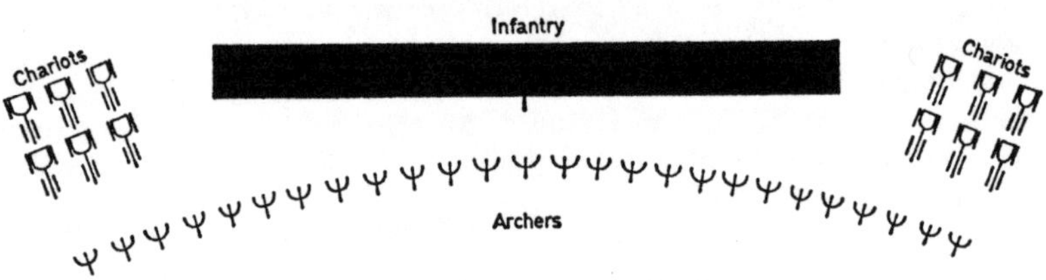


An Egyptian battle formation

## STYLE OF FIGHTING

At first, this was much the same as the Sumerian style. Then men began to be formed into units held together with discipline and given a certain amount of training. In battle, spearmen stayed together in large groups, lightly armed archers and slingers swarmed out in front whilst the nobles, mounted or in chariots, surged around on the flanks. Manoeuvre was usually accidental as the armies advanced towards each other, both keeping up a harassing fire until the chariots and cavalry charged. Then the light troops drifted to flank and rear through ill-defined gaps between the large groups of infantry. Sometimes the chariots and the horsemen so frightened the enemy that they turned and ran so that the battle turned into a murderous chase. Otherwise, both masses clashed together in a huge mêlée with the line swaying backward and forward until one side suddenly broke—then the chase began. The Battle of Megiddo in 1469 B.C. is the first recorded battle of history in which Thothmes III certainly had his army under sufficient control to permit tactics that gave the Egyptians overwhelming victory. At this time the Egyptian troops appear to be well disciplined and drilled.

*Fighting Assessment Chart* ANCIENT EGYPTIANS

| *Factor* | *Rating* | *Points* |
|---|---|---|
| 1. Ability to assault in controlled order | Little | 1 |
| 2. Bravery | Reasonable | 2 |
| 3. Control | Reasonable | 2 |
| 4. Defensive ability | Little | 1 |
| 5. Desperation | Not applicable | – |
| 6. Discipline | Reasonable | 2 |
| 7. Ferocity | Not applicable | – |
| 8. Manoeuvrability | Little | 1 |
| 9. Mobility | Reasonable | 2 |
| 10. Movement, speed of | Reasonable | 2 |
| 11. Order | Little | 1 |
| 12. Tactical Sense | Little | 1 |
| 13. Training and Drilling | Reasonable | 2 |
| 14. United action, capability for | Little | 2 |
| 15. Weapon Handling | Reasonable | 2 |
| 16. Demoralisation, likelihood of | Unlikely | 2 |
| 17. Impetuous Action, likelihood of | Little | 2 |
| 18. Panic, likelihood of | Unlikely | 2 |
| 19. Reluctance to take definite action | Little | 2 |
| 20. Unscheduled Manoeuvres, likelihood of | Unlikely | 2 |
| | *Fighting Assessment* | 31 |

# THE ASSYRIANS

At about the same time as the Egyptians flourished, the new nation of Assyria was coming to the fore in the East, inheriting and extending the ancient civilisations of Sumer and Babylon in the fertile country near the head waters of the Tigris. For nearly 300 years (from 884 to 610 B.C.) the Assyrians pillaged and plundered their unfortunate neighbours from the Tigris–Euphrates to the Nile. Making a deliberate policy of ferocity, the Assyrians deported whole populations and were probably the first race on earth to transplant captured peoples from their native land. Hardy and warlike, the Assyrians stand out as the greatest pre-Roman example of a military state ruled by a king who was also a competent soldier backed by the most ferocious and best equipped army the world had seen up to that time. They showed a great thoroughness in planning, exemplified by a system of excellent roads and staging posts which gave them a great mobility.

At first, the chariot was the main Assyrian weapon and its

crews formed the elite of the army. The chariots were of two types—a light two-man vehicle bearing an archer and a charioteer and a heavier chariot carrying four men—archer, charioteer and two shield bearers. These vehicles had no springs which must have made the firing of a bow from them a difficult business so that it is quite possible that at times the chariot halted and the bowmen either fired from a stationary platform or descended and fired their missiles from the ground. Egyptian chariots had a 'floor' of crossed leather thongs, thereby giving some measure of 'springing' for the occupants. It could be that the Assyrians used something similar—they had been in contact with the Egyptians for some long time. While the archers were in action their attendant shield bearer warded off enemy missiles. Reliefs of the time show the chariot horses to be wearing a type of barding or protective armour, probably of some heavy quilted material. The chariot crew wore short sleeveless coats of scale armour made of small overlapping plates sewn on to a cloth backing. They carried short swords slung by a strap on the left side and small round shields; they wore the typical high peaked helmet with small lappets over the ears.

The Assyrians placed a lot of reliance on fire-power and the archer figures prominently in their warlike activities. There seems to have been two types—a lightly armed soldier who fought bare-headed and barefooted clad only in a kilt-type garment and the heavily armoured bowman who wore a hip-length coat of scale armour over a long fringed robe which reached to the ankles or sometimes a knee-length tunic below which were leg coverings apparently made of some form of chain mail. The helmets were the same conical type as worn by the charioteers. Shield bearers, similarly armed and carrying a spear or sword, protected the heavy archers, in some cases bearing wicker mantlets taller than a man and wide enough to cover two or three men. Even the slingers, who in other armies were always the lightest equipped troops, wore armour. This indicates that the Assyrian tactics relied on the holding of advance positions by heavily protected missile men. The Assyrian spearmen were divided into light and heavy units, wearing armour and helmets similar to those of the corresponding class of archers although the light spearman is sometimes shown wearing a helmet with a crest instead of the high peak.

Perhaps the greatest Assyrian contribution to the art of war was the soldier on horseback, although he does not seem to have made his appearance until the time of Sargon II and his son Sennacherib (*c.* 722–681 B.C.). The great conqueror Tiglath-pileser I (*c.* 1130–1110 B.C.) does not seem to have used cavalry and the sculptures of the time of Ashurnasirpal (*c.* 883–858 B.C.) show a few horsemen but far more chariots. The Assyrians had horse archers, armed with sword and a small bow firing three-foot arrows, and lancers with swords and ten-foot lances although, without saddle or stirrups, it is probable that these men were not particularly efficient. The Scythians were allies of the Assyrians for a time, and they could have learned the use of horse archers from them. Instead of shields, the horsemen wore flexible waist-length jackets made of metal scales and lace-up boots; sometimes their horses wore a protective covering. Reliefs and sculptures show that the early cavalry horses had equipment almost identical to that of the chariot horses and shows the archers being attended by an unarmed man, which leads to the theory that perhaps these archers were originally charioteers who had unhitched the horses and ridden off when the ground was unsuitable for the use of their chariots.

The Assyrians made a speciality of siege tactics, employing movable towers with battering rams at the base and archers at the top, to wheel up against the wall.

Together with the Egyptians and the Babylonians, the Assyrians were the first people to make use of trained armies and under the leadership of perhaps their greatest king, Ashurbanipal, Egypt was subdued, Susiana conquered and Babylon held in a firm grip. Assyrian armies fought their way deep into Armenia and across the rugged Taurus Mountains into Asia Minor. The power of the Assyrians was finally broken by the Medes who besieged Nineveh in 606 B.C. and with its capture the Assyrian Empire ended.

## STYLE OF FIGHTING

Probably the best drilled and disciplined army until the Romans, the Assyrians were extremely ferocious and used terror as a tactical factor. In battle, their highly organised archers caused confusion in the enemy ranks as preparation to a well controlled chariot and cavalry charge. Their large masses of spearmen, more manoeuvrable than similar formations of the period, finished the battle by a slow-moving but relentless advance.

Use TERROR as a tactical factor.

*Fighting Assessment Chart:* ASSYRIANS

| | *Factor* | *Rating* | *Points* |
|---|---|---|---|
| 1. | Ability to assault in controlled order | First Class | 3 |
| 2. | Bravery | First Class | 3 |
| 3. | Control | First Class | 3 |
| 4. | Defensive ability | Reasonable | 2 |
| 5. | Desperation | Not applicable | – |
| 6. | Discipline | First Class | 3 |
| 7. | Ferocity | First Class | 3 |
| 8. | Manoeuvrability | First Class | 3 |
| 9. | Mobility | Reasonable | 2 |
| 10. | Movement, speed of | Reasonable | 2 |
| 11. | Order | First Class | 3 |
| 12. | Tactical Sense | Reasonable | 2 |
| 13. | Training and Drilling | First Class | 3 |
| 14. | United action, capability for | Reasonable | 2 |
| 15. | Weapon Handling | First Class | 3 |
| 16. | Demoralisation, likelihood of | None | 3 |
| 17. | Impetuous Action, likelihood of | None | 3 |
| 18. | Panic, likelihood of | None | 3 |
| 19. | Reluctance to take definite action | None | 3 |
| 20. | Unscheduled Manoeuvres, likelihood of | None | 3 |
| | | *Fighting Assessment* | 52 |

# THE PERSIANS

Most of the technical knowledge and military organisation of the Assyrians was inherited by the Persians but they lacked the more ferocious aspects. By the 6th century B.C. the Persians had extended the Assyrian Empire so that it reached from the borders of India to Egypt and the Aegean. The various tribes of the Persians were first united under the leadership of Cyrus who defeated his Median rulers in 560 B.C. and then took over the lands and treasures of the trading empire of Lydia. In the battle fought to achieve this aim, it is said that Cyrus formed his baggage camels into a fighting unit to challenge the Lydian superiority in cavalry. As might be expected, the scent of the camels frightened the Lydian horses so that their riders were forced to fight on foot, a disadvantage which gave victory to the Persian infantry, armed with short spears and small round shields.

The Greek colonies on the Asiatic coasts, formerly ruled by the King of Lydia, now found themselves in direct contact with the

Persians. Disunited and displaying a blind reluctance to present a common front to the foe, they fell one by one to Cyrus's generals. For some years the Greek States provided ships and men for the Persian armies but during the reign of Darius, successor to Cyrus's son Cambyses, a series of revolts overthrew the despotic rulers of the city-states who had been maintained in their position by Persian arms. Darius's army was established on a divisional basis, 10,000 men to a division, with a core of lightly armed bowmen backed by large forces of cavalry and 10,000 'Immortals' of the king's bodyguard. With the exception of the latter force who were all Persian, many of these soldiers were auxiliaries drawn from the far corners of the vast Persian Empire and armed and equipped after the fashion of their native lands. There were Caspian cavalry wearing goat-skin capes and carrying scimitars and bows made of cane; Median cavalry in sleeved jackets covered with iron scales, carrying longbows and short spears and shields made of plaited reeds. The nomadic Sargatians were armed only with lassos and daggers, whilst cloaked Arabs rode camels. It is said that many of these auxiliaries had to be forced into battle with whips wielded by their sergeants and officers. The Persian naval forces commanded the sea and supported expeditions carried out near the coast. Siege weapons did not seem to have been used by the Persians who captured cities by mining or other means not requiring siege engines.

The basic Persian tactic consisted of galling the enemy with arrows and then launching cavalry against him. The Persian line of battle had their infantry in the centre, archers in front and cavalry on the wings. Preceded by light archers and slingers acting in a skirmishing role, the infantry moved forward until within bowshot when the heavy archers set down their wicker mantlets and commenced to pour clouds of arrows into the enemy. When they were deemed to have been sufficiently softened up, the cavalry charged in decisively from the wings.

These tactics succeeded with bowmen like the Medes, mounted lancers like the Lydians, heavy infantry like the Babylonians and the Egyptians but against the Greeks they came to grief. In the first place, the Persian bows were not powerful enough to put up an arrow storm sufficient to break up the advance of the Greek hoplites, who made short work of the Persian footmen, with their short swords and small shields, if they got to close quarters. Aeschylus described the Graeco–Persian Wars as a contest between the bow and the spear in which the spear proved itself the superior weapon. The second factor that made for Greek triumph was the comparative lack of success of the Persian cavalry who charged in loose order whilst employing javelins and bows instead of attaining a shock-value by charging in knee-to-knee ranks. Although they did not treat the Persian cavalry lightly, the Greeks were usually able to counter them by stationing light infantry on the flanks of their phalanx.

Their defeat at Marathon (490 B.C.) shook the confidence of the Persian soldiers in themselves and their leaders as it was a severe check to their unblemished record of success in battles and conquest. The Persian victory at Thermophylae in 480 B.C. was costly, being followed by a Greek sea victory at Salamis and another Persian defeat at Plataea in 479 B.C., shattering once and for all the legend of Persian invincibility and the dream of a Persian conquest of Greece.

## STYLE OF FIGHTING

Under Cyrus and later Darius, the Persian military system was of a high order. Versatile and adept in suiting their methods of warfare to all conditions of terrain, the Persians avoided close-quarters infantry combat until their foot archers were in the front and their horse archers in flank and rear had confused and disorganised the enemy. The Persians, like most Asian armies, knew how to employ cavalry but were never able to fully co-ordinate their horsemen with their infantry—even so, they managed this better than their opponents!

At the battle of Thymbra in 546 B.C. Cyrus, badly outnumbered, formed the first recorded tactical deviation from the normal parallel order of combat. The bulk of his army formed up in depth, in five relatively short lines, their flanks covered by chariots, cavalry, his best infantry and an improvised Camel Corps, facing outward, perpendicular to the front. The Lydian army, wheeling inwards to envelop him, allowed gaps to appear at the hinges of the wheeling wings. Disordered by overhead fire from the Persian archers stationed within his square formation, the Lydians had their disorganised wing smashed by Cyrus's flank units and the rout was completed by the Persian cavalry.

The Persians used auxiliary troops from subject peoples, and sometimes Greek mercenaries. Although the auxiliaries were surprisingly loyal, the fact that they were subject peoples should be taken into account. The mercenaries can be considered as suggested elsewhere in this book.

*Fighting Assessment Chart:* PERSIANS

| | *Factor* | *Rating* | *Points* |
|---|---|---|---|
| 1. | Ability to assault in controlled order | Reasonable | 2 |
| 2. | Bravery | Reasonable | 2 |
| 3. | Control | Reasonable | 2 |
| 4. | Defensive ability | Average | 1 |
| 5. | Desperation | Not applicable | – |
| 6. | Discipline | Reasonable | 2 |
| 7. | Ferocity | Not applicable | – |
| 8. | Manoeuvrability | Reasonable | 2 |
| 9. | Mobility | Reasonable | 2 |
| 10. | Movement, speed of | Reasonable | 2 |
| 11. | Order | Reasonable | 2 |
| 12. | Tactical Sense | Average | 1 |
| 13. | Training and Drilling | First Class | 3 |
| 14. | United action, capability for | Reasonable | 2 |
| 15. | Weapon Handling | First Class | 3 |
| 16. | Demoralisation, likelihood of | Average | 1 |
| 17. | Impetuous Action, likelihood of | Average | 1 |
| 18. | Panic, likelihood of | Average | 1 |
| 19. | Reluctance to take definite action | Average | 1 |
| 20. | Unscheduled Manoeuvres, likelihood of | Average | 1 |
| | | *Fighting Assessment* | 31 |

# THE GREEKS

Around 2000 B.C. a new civilisation had grown up in the Aegean. Originally created by the Cretans, it flourished under the Myceneans or Greeks of the mainland, whom Homer later immortalised in his epics. The Achaeans, northern invaders, conquered and finally mingled with the original inhabitants of this Minoan or Aegean civilisation. From the mixing of the two races and cultures came Homer's 'Heroic Age', placed from about 1500 B.C. to 1100 or 1000 B.C. It was probably during this period that the struggle occurred between the Achaeans with their allies and the rulers of the lands around Troy, immortalised in Homer's poems *The Iliad* and *The Odyssey*. Despite the beauty of this poetry, warfare of the time was almost devoid of planned manoeuvring or real leadership. The 'Heroes' rushed into battle in their chariots followed by the infantry 'shield to shield, shoulder to shoulder' to take part in a vast sprawling mêlée where spears, rocks, swords and daggers took their close-quarters toll.

It has been said that the Achaeans rather resembled the bloodthirsty Vikings who came later in history. Their tactics were looser and more individual than those of the Egyptians, for example.

The most interesting contest of the period was the Trojan War which lasted from about 1194 to 1184 B.C., when Agamemnon led his armies of chariots and infantrymen into Asia Minor to put an end to the menace of the Trojans who had established themselves in Troy, their fortress kingdom. Homer tells of chariot lines meeting each other with great shocks, vast mêlées between onrushing infantrymen and hails of javelins as soon as chariots came within range, when the occupants leapt from their carts to attack the enemy on foot.

Between 1200 and 1000 B.C. the Mediterranean world fell into chaos as the Mycenaean and Hittite civilisation crumbled before invading hordes. It was the watershed between the ages of bronze and of iron, the bronze being no match for the slash of the iron blades carried by the invaders.

Little is known of the troubled centuries that followed until the 8th century B.C. when the Greek city-states emerged. Consisting of a walled town surrounded by outlying farms and villages, a city-state was often within a few hours' march or even in sight of another similar town with which it was at war. The armies of the city-states were made up of citizen-soldiers, armed and marshalled according to their means. It is doubtful whether any Greek city except Athens could ever put more than 20,000 men into the field, including all sound males between sixteen and sixty.

The chariot had gone out of use and the Greek method of warfare was based on the heavily armed and armoured spearman—the hoplite, a well-to-do citizen who could afford arms and armour, his equipment was practically standardised throughout the Greek world. The hoplite wore a crested metal helmet and a metal cuirass hinged at one side and suspended from the shoulders by a thick leather strap. His legs were protected by greaves long enough to cover the knee and conforming to the shape of the leg. It was a protective armour that well suited the athletic Greek. His main weapon was a heavy thrusting spear made ten feet long and a leaf-shaped two-edged sword. They carried a round, deeply dished shield some three feet in diameter held by a strap through which passed the left forearm.

Each hoplite was accompanied by at least one attendant who acted as a shield bearer and served as a lightly armed soldier when fighting commenced. The citizens who could not afford to buy the armour of a hoplite served as auxiliaries—javelin men, archers and slingers. They were variously equipped but the javelin man normally carried a round shield smaller and lighter than the hoplites and his helmet was a small casque or skull cap perhaps of leather. Some of the lighter armed troops wore leather armour and cuirasses made of many folds of linen glued or stitched together.

Before the days of the hoplites, the Greek method of fighting was for their 'Heroes' or champions to go forward and fight a personal duel with the champion of the opposing force. Now, the heavily armoured spearmen, trained to work together, went forward in a phalanx where the safety of the whole array depended on each man supporting his neighbour. This massed array of spears was an ideal formation to suit the limited training and capabilities of the citizen-soldier, providing a formation reasonably easy to handle with a minimum of training whilst it gave each member of the phalanx a high degree of moral and physical support. This tightly packed rectangular formation, sometimes eight ranks deep by several hundred files long, was not very flexible—its main function being to advance straight to the front in a manner that was difficult, if not impossible, to halt or to make change its direction. On the flat ground usually chosen for their battles, the phalanxes of both sides formed up and at first advanced slowly, then at the run encouraged by loud shouts. The fighting was done by the front two ranks, the files of the formation behind them being fed in as the front rank men fell. Perhaps because of greater impact or a denser formation, one side would begin to gain an advantage and the weaker side would begin to give ground until the whole mass broke up and ran, the victors prodding and stabbing at their retreating backs. Because of the lack of cavalry, there was little pursuit and such a clear-cut victory often ended the war at the first attempt. The flanks of the phalanx were extremely vulnerable and if attacked the flanking files were forced to face outward, so halting any forward movement. The rear of the phalanx was its most sensitive point and it was not difficult for the attackers to get round these. This is assuming that the attackers had suitable light infantry or cavalry otherwise in the normal hoplite battle it would be difficult.

The lighter armed troops did not do very much in the battle, lacking both discipline, training and adequate arms and equipment. Except for Thessaly, a city-state which relied mainly on armoured cavalry, horsemen did not play a great part in these battles and only the very wealthy, who could afford both horse and armour, took their place on one or other of the wings of the main body to drive off the opposing slingers, archers and javelin men or to engage in a furious mêlée with the cavalry of the opposing side. The standard cavalry weapon was the short spear and javelins, with the sword as close-quarters weapon. Occasionally mounted archers were present. The Athenians had an elite corps of young and wealthy cavalrymen, sometimes reaching a strength of over a thousand, whilst the Boeotians, to the north, made considerable use of cavalry particularly under Epaminondas at Leuctra and the second battle of Mantinea.

There were several types of Greek cavalry, the heaviest being the cataphracti, copied from the Persians, who wore helmets, cuirasses and thigh pieces and whose horses were protected by head and breast armour. But, generally speaking, Greek horses were too small to be armoured effectively—a few nobles might have some kind of front protection for the horses, but no significant numbers. Then there were the medium cavalry, wearing lighter armour or none at all and with unprotected horses and finally the Tarentine or irregulars variously armed with bows and javelins.

The Greek hoplites won many victories against the Persians, notably at Marathon in 490 B.C. and Plataea in 479 B.C., when Darius and his son Xerxes invaded Greece. The twenty-seven years for which the Peloponnesian War dragged on brought an enforced change in the system of citizen-soldiers, who went back to their business whilst the state used the professional (mercenary) heavy armed spearman, a better trained man than the average citizen-soldier. It was in the quality of the professional light infantryman, the peltasts, that the greatest change was seen. Taking the name from the small shield that they carried, they had the discipline and training to successfully fight in open order. Some of the city-states, particularly the wilder ones in the mountainous north, produced naturally efficient light infantry. In 429 B.C. an Athenian expedition against the Chalcidians met disaster at the hands of such troops and only three years later the Athenian Demosthenes was similarly defeated by the swift-footed and light-armed Aetolians. In later years the Athenians made great improvements in the training and equipment of their peltasts, providing them with light armour, larger shields and longer javelins and swords. In the Corinthian War of 390 B.C., the lightly armed man proved that when correctly handled he was a serious threat to the heavy infantryman.

The relative military values and fighting qualities of all the Greek city-states is not known although there is sufficient knowledge available about Sparta, Athens and Thebes to give a good idea of the manner in which they conducted their warfare. The necessary alliances between these city-states form a bewildering and constantly changing pattern of Greek history so that it was a period of wars, revolts and bloody civil strife. Ironically, although bound together by ties of race, the Greeks spent most of their time internally destroying their civilisation until a stronger outside power destroyed them.

Above all of the city-states, the people of Sparta were probably the most remarkable inhabitants of the ancient world, renowned for their physical prowess and dogged courage. So well known was their strength and endurance that the very appearance of a Spartan force in the field was often enough to ensure victory. The Spartan was more than a match for even the professional soldier because he devoted his entire life to military training and he had an immensely high morale brought about by single-minded patriotism and the superiority encouraged by an almost unbroken succession of victories. Small in size, and partly because of the peculiar structure of the State, Sparta was never able to put more than 5,000 hoplites into the field, backed by perhaps the same number of perioeci (conquered people conscripted into service) and a force of helots or serfs who were conscripted to serve as squires to the Spartan warriors and to act as lightly armed troops.

At the battle of Mantinea, Thucydides wrote that there were seven morai (A mora was a company of 512 men) in the field . . . in each company there were four pentecostyes (a platoon of 128 men) and in the pentecosty four enomoties (a section of 32 men). The first rank of the Enomoty was composed of four soldiers . . . generally ranged eight deep . . . the first rank along the whole line consisted of 448 men. There was also a king's bodyguard of 300 picked knights who fought on foot. Cavalry were introduced into the Spartan army in about 424 B.C. and consisted of 6 morai, each of about 100 horses, divided into two squadrons.

The Spartan morais' equipment was similar to that of any other Greek hoplite although red tunics have been mentioned as being a distinctive part of his dress. In about 230 B.C., the Spartans adopted the sarissa (a pike some 16 to 24 feet in length).

Plutarch reveals an interesting facet of the Spartan way of war when he tells how they never pursued an enemy when they were well assured of victory, thinking it unworthy of a Greek to kill an enemy who had given up resistance. This was a shrewd rather than a magnanimous tactic because their enemies, conscious that the Spartans killed only those who resisted, were more likely to save their lives by fleeing at the first sign of a battle turning against them.

The long military supremacy of Sparta was basically due to their sheer professionalism, exemplified by the manner in which the Spartan phalanx advanced in a slower and more menacing fashion than similar formations of other Greek states.

Spartans won many great victories on hard-fought fields but their finest hour was probably at Plataea in 479 B.C. when 75,000 men from twenty city-states banded together to fight off about 100,000 Persians under Mardonius. Sundry tactics and manoeuvres caused almost the entire Persian strength to fall upon the Spartan force of about 5,000 hoplites with similar numbers of attendant perioeci and helots, under Pausanias. At first held by the Persian cavalry, the Spartans then had to face a swarm of Persian archers who poured volley after volley into their ranks. Unable to come to grips, the Spartans seemed to be in a disastrous position but, in their eagerness to press forward and crush their dreaded foes, the Persian infantry piled up upon one another until they lost all power of manoeuvre. Seeing this, Pausanias gave the order to charge and the massed phalanx crashed into the disordered mob of light-armed Persians who were so bunched together that they could not retreat. In the desperate fight that followed, the Persian leader Mardonius was killed and his men withdrew in a headlong flight from the field.

Although a defeat the Spartan resistance at Thermopylae, in 480 B.C. when 300 Spartans died to the last man, has come down in history as an almost legendary tale of heroism.

Like all other great powers, Sparta declined and the victory of Antigonus over Cleomenes, the Spartan king, at Sellasia in 221 B.C. ended Sparta's long reign.

### *Athens*

The antithesis of Sparta in her sophistication and joy of living, Athens rose to her peak during these years of strife by becoming a maritime power and acknowledging the superiority of Sparta on land. But although her forces marched as allies or against the other city-states, her hoplites showed their true fighting qualities as warriors and marines on the decks of her warships.

In time of war every Athenian citizen was liable for service—the poorer classes as light troops and the wealthier classes as cavalry or hoplites. In particular, Athens developed missile infantry, discovering in certain circumstances the slow-moving Spartan phalanx could be outwitted by using lightly armed infantry and cavalry to strike, preferably unexpectedly, on the flank, retreat and then strike again. Another tactic was to draw the phalanx into wooded mountainous or broken country so hoping that its tight formation and crushing power would be broken—on the open plain the Spartans were almost invincible. At the beginning of the Peloponnesian War (431 to 404 B.C.) Athens and Sparta clashed head on, the first victory going to Sparta thanks to its phalanx and superior leadership. Later in the war, the archers of Athens took such a heavy toll of Sparta's infantry that the Spartans in their return raised a body of archers. Athens then created an army of peltasts (slingers)—light infantry trained to manoeuvre rapidly so that it was almost impossible for the Spartan phalanx to pin them down. At this stage an Athenian expedition was often formed of about 600 archers and peltasts to 1,000 hoplites.

Athens had a strong cavalry force divided into phylai (squadrons)

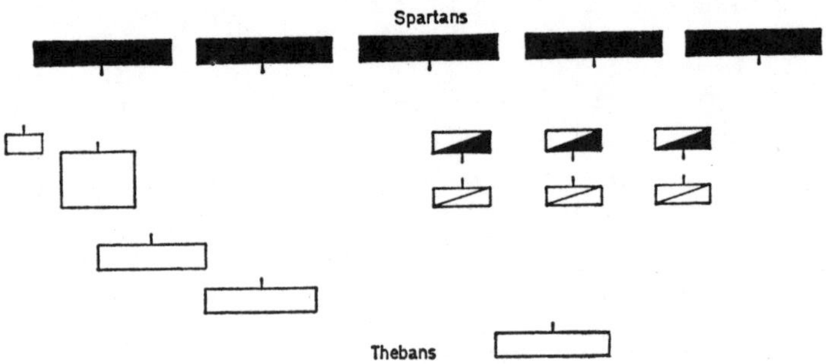


Leuctra 371 B.C.

of 100 or more horse each; there were ten such squadrons under the command of two cavalry generals (hipparchoi).

*Thebes*

When they were not fighting the Persians, the Greeks fought each other, using the same weapons and the same military formations so that battles were decided by numbers or by the fluctuations and chances of war. It was not until 371 B.C. that Epaminondas, a Theban general, revolutionised the time-honoured tactics in a manner that, with hindsight, seems very simple and logical. A characteristic of the phalanx was that it tended always to bear to the right because each man carried his shield on his left arm and tried instinctively to bring his unprotected right side under his right-hand comrade's shield. The almost subconscious right oblique movement of two phalanxes charging each other usually resulted in the right wing of each formation simultaneously out-flanking and defeating the enemy left wing so that the battle was decided by a confused mêlée when the victorious right wings wheeled and attacked each other. For this reason, the best troops were always placed on the right of the phalanx. Epaminondas reversed this order by stationing his best fighting division, the Theban Sacred Band, on the *left* wing, strengthening its thrusting power by forming it in a column fifty ranks deep. At Leuctra in 371 B.C., a weak force of Spartan cavalry was soon driven from the field and then the Spartan right-wing phalanx marched forward in what had hitherto been their irresistible advance. At the same time the Thebans moved forward in echelon, their strong left wing in advance and their weaker right wing refused (held back). The twelve-deep Spartan phalanx could not resist the impact and steady thrust of the much deeper Theban formation and in the struggle that followed they were shattered and their king Cleombrutus killed. In this first recorded instance of the 'oblique order' (to be successfully practised nearly 2,000 years later by Frederick the Great) the Thebans killed 400 Spartans in a defeat which shocked Sparta and amazed the Greek world.

The Theban tactics that had succeeded at Leuctra again won the day at Mantineaa in 362 B.C. But in the second of these battles Epaminondas was killed and the brief supremacy of Thebes, due in great part to the efforts of this fine soldier, came to an abrupt end.

## STYLE OF FIGHTING

Except in the northern flatter regions of Thessaly and Macedonia, the Greeks neglected cavalry and relied on the steadily improving infantry phalanx.

In early times, Greek warfare was notable for its 'Heroes'—champions who went forward and engaged their counterparts from the enemy's ranks in individual combat. That part of the battle concluded, both sides clashed together in a formless mêlée where each man fought for himself.

Later, the Greek formation was often a long solid line with narrow avenues for the passage of light troops. In combat formation, the phalanx was capable of limited manoeuvres but movement over rough ground could create dangerous gaps in its ranks.

The general tactics of the Graeco-Persian Wars basically consisted of attempting to force the enemy to enter battle on unfavourable ground, or with only a portion of his available forces, or to out-flank him. The armies lacked that essential control which prevented the exposure of a marching flank or the creation of dangerous gaps in lines by anything other than leisured manoeuvres whilst still out of missile and mêlée range. Approaching each other in parallel lines, the armies formed up with the infantry in the centre and cavalry on the wings, the advance screened by clouds of slingers and bowmen who would melt away around the flanks or through narrow intervals left in the ranks immediately before the main line of the opposing armies clashed together. Occasionally one side stood on the defensive whilst another attacked, otherwise about the only variation in battle order was for cavalry and chariots to be in front of the main battle formation instead of on the flanks.

*Fighting Assessment Chart:* GREEKS

| | *Factor* | *Rating* | *Points* |
|---|---|---|---|
| 1. | Ability to assault in controlled order | First Class | 3 |
| 2. | Bravery | Reasonable | 2 |
| 3. | Control | Reasonable | 2 |
| 4. | Defensive ability | Reasonable | 2 |
| 5. | Desperation | Not applicable | – |
| 6. | Discipline | Reasonable | 2 |
| 7. | Ferocity | Not applicable | – |
| 8. | Manoeuvrability | Reasonable | 2 |
| 9. | Mobility | Reasonable | 2 |
| 10. | Movement, speed of | Reasonable | 2 |
| 11. | Order | Reasonable | 2 |
| 12. | Tactical Sense | Reasonable | 2 |
| 13. | Training and Drilling | Reasonable | 2 |
| 14. | United action, capability for | Reasonable | 2 |
| 15. | Weapon Handling | Reasonable | 2 |
| 16. | Demoralisation, likelihood of | Average | 1 |
| 17. | Impetuous Action, likelihood of | Average | 1 |
| 18. | Panic, likelihood of | Average | 1 |
| 19. | Reluctance to take definite action | Average | 1 |
| 20. | Unscheduled Manoeuvres, likelihood of | Average | 1 |
| | | *Fighting Assessment* | 32 |

# THE MACEDONIANS

The Macedonians, whose kingdom lay to the north of Greece, were a warlike race with Greek stock and traditions but lacking their characteristic culture. Philip, who unified the Macedonians, had spent some years as a hostage in Thebes and had learned much from Epaminondas. On his return, he improved upon the Theban phalanx, increasing its mobility by thinning its ranks to 16 and by lengthening the pikes (sarissa) to 16, 18 and even 24 feet so that the weapons of men as far back as the fifth rank projected beyond the soldiers of the front rank. To cope with this two-handed weapon, the shield was reduced in size and struck to the left arm, leaving the hand free. With these exceptions, the arms and equipment of the Macedonian peasant-soldier were not very different from that of the regular Greek hoplite. The usual proportion of cavalry to infantry in most Greek states was about 1 to 12 or 1 to 16, but Philip, believing in the riding habits of his 'farmers', had a higher proportion of about 1 to 7.

Inevitably, Macedonia eventually came into conflict with the other Greek states, and at the battle of Chaeronea in 338 B.C. Philip defeated a combination of the Athenian and the Thebans. Although this victory gave him control of Greece, Philip never achieved his dream of uniting the Greek city-states for an attack on Persia. Philip was murdered in 336 B.C. and his splendid army came under control of his son Alexander, a man destined whilst still young to become one of the greatest generals the world has known. By breaking the phalanx into smaller units Alexander turned it into a spear-hedged mobile base rather than a juggernaut of moving spear points. But the real striking force of the Macedonians lay in the shock power of their heavy cavalry charging in formation at speed. They were formed of eight squadrons—the Companions on the right wing and Thessalians on the left, both being supported and strengthened by light cavalry and light infantry. Alexander also devised a new class of highly trained foot soldier—the hypaspists—who were a cross between the heavy pikemen of the phalanx and the light-armed peltast. They wore armour and were armed with a shorter spear so that they formed a link between the phalanx and the cavalry. However, there is still doubt as to the actual equipment of the hypaspists. A picked body of these hypaspists constituted the Royal Foot Guards (the Agema).

In battle the Macedonian phalanxes advanced in echelon so that the right division struck the enemy first and 'fixed' him. Then the Companions, often under Alexander himself, attacked the enemy's left supported by the hypaspists. If the enemy attempted to outflank the left of the Macedonian phalanx then the Thessalian cavalry, together with light cavalry and light infantry, opposed them whilst a similar screen guarded the right of the Companions and moved up to engage the enemy left with the heavy cavalry attack succeeded. It was a system that required mutual support of all arms arising from a sense of confidence in each other and in their leader.

Alexander took his armies into Persia, gaining victories at Granicus in 334 B.C. and at the Issus in 333 B.C., but the decisive battle of the war was fought at Arbela in 331 B.C. Here, Darius had more than 200 scythed chariots but they were well handled by the Macedonian archers and javelin men who shot down the drivers or the outer horses causing them to fall entangled in their traces, so immobilising the vehicles. The Macedonian cavalry then swept in to cut down, with the lance or long sword, the helpless charioteers. There was only one unit formed of lancers in the Macedonian army, cavalry armed with the sarissa, but the rest would normally have the 'standard' short spear, sword, and javelins in some cases. It was the training and discipline, combined with Alexander's own leadership, and timing, that made the Macedonian cavalry so formidable—even then, it is doubtful if they would be able to break steady hoplite infantry in a frontal charge.

Travelling as far as the Punjab in India, Alexander's army fought a fierce battle with Porus, an Indian ruler. Here the Macedonians had their first encounter with elephants, considered a terrible weapon of war whose very appearance struck terror in the hearts of the enemy. The elephants' hide was almost impervious to missiles, it had formidable tusks and a trunk which could fling horses and men into the air. On its broad back it carried a little wooden fort packed with missile-men. By sheer weight alone, the animal was a bulldozer that crushed anything in its path. But the military value of elephants was always exaggerated and Alexander's infantry quickly learned the way to deal with them—it had to be the infantry because the horses of the cavalry would not face the elephants. Maddened by the spears and arrows of the Macedonian light infantry and rarely handled very expertly, the elephants threw even their own ranks into great confusion.

The Graeco-Macedonian armies reached their peak under Alexander, who had the ability to arouse great loyalty amongst his followers, together with a supreme skill in freely manoeuvring his well trained army in small units—one of the great developments of the art of warfare to come out of this period.

## STYLE OF FIGHTING

Philip of Macedon is said to have created the first scientifically organised military force of history when he used heavy missile engines in co-ordination with the field operations of infantry and cavalry—it was the forerunner of field artillery.

Alexander used an oblique order of battle, echeloned back from the right-flank cavalry spearhead or with the hypaspists on the right flank of the phalanx to provide a flexible hinge between the fast-moving cavalry and the relatively slow hoplites. Flanks and rear were protected by light infantry, peltasts drawn up in a line approximately eight men deep behind the phalanx and psiloi (servants and foragers for the heavy infantry) formed a skirmish line in front; they were armed with bows, javelins, darts and slings. Besides being a base of manoeuvre for the shock action of his cavalry, the heavy infantry phalanx was a highly mobile formation, capable of attacking at the run whilst retaining formation so as to hit the disordered enemy ranks before they could recover from the impact of the early cavalry attack (see page 45, Dupuy).

Alexander the Great was a military genius, capable of adjusting his tactics and tactical formations to suit prevailing conditions. His tactical skill in mountain warfare and against irregular forces was of the highest order and he must be considered to be one of the three greatest military leaders in history. His father Philip of Macedon was also an extremely good general and could be classed as an 'Above-average' commander.

*Fighting Assessment Chart:* MACEDONIANS

| *Factor* | *Rating* | *Points* |
|---|---|---|
| 1. Ability to assault in controlled order | First Class | 3 |
| 2. Bravery | First Class | 3 |
| 3. Control | First Class | 3 |
| 4. Defensive ability | Reasonable | 2 |
| 5. Desperation | Not applicable | – |
| 6. Discipline | Reasonable | 2 |
| 7. Ferocity | Not applicable | – |
| 8. Manoeuvrability | First Class | 3 |
| 9. Mobility | First Class | 3 |
| 10. Movement, speed of | First Class | 3 |
| 11. Order | First Class | 3 |
| 12. Tactical Sense | Reasonable | 2 |
| 13. Training and Drilling | First Class | 3 |
| 14. United action, capability for | First Class | 3 |
| 15. Weapon Handling | Reasonable | 2 |
| 16. Demoralisation, likelihood of | Rarely | 2 |
| 17. Impetuous Action, likelihood of | Average | 1 |
| 18. Panic, likelihood of | Rarely | 2 |
| 19. Reluctance to take definite action | Unlikely | 3 |
| 20. Unscheduled Manoeuvres, likelihood of | Average | 1 |
| | *Fighting Assessment* | 44 |

# THE ROMANS

At a time when the Greeks in their wars between the various city-states were employing the phalanx and a reasonable degree of military knowledge, the Roman farmer-soldiers were making war by ambush or sortie in relatively undisciplined bands that returned to their fields and harvest at the first sign of a lull in the fighting. In those early days, Rome, as a city-state that had to defend itself, found it necessary to enrol all males of fighting age into a state militia. In this the first of the several phases in the development of the Roman army, the militia was divided into units according to the financial ability of its soldiers to equip themselves. This meant that the rich fought with arms, armour and perhaps horses whilst the poor formed the bulk of the lighter armed and unarmoured infantrymen. The Romans got their basic idea of the legion from their Greek neighbours and their early tactics differed little if at all from those normally employed by the Greeks. Like the Greek phalanx, the legion was the military unit of a military minded state and consisted of 3,000 to 4,000 citizens and property-owners arranged in three divisions according to age and experience, the veterans backing up the younger soldiers, with light infantry and cavalry on the wings. There is no evidence that the Roman soldier of that time was in any way superior to the Greek and there would probably have been little to choose between them had these similarly armed and equipped men met on the field of battle.

It is known that in the days of Servius Tullius (*c.* 580 B.C.) the army totalled nearly 20,000 men made up of 18 centuries of cavalry; 80 first-class fully armed centuries; 10 second-class centuries carrying wooden shields; 10 third-class centuries without greaves; 10 fourth-class centuries without breast-plate or helmet and armed with a wooden shield, sword and spear and 10 centuries of the poorer citizens armed with darts and slings. Their formation was the familiar phalanx with the best equipped spearmen in front and cavalry and light troops on the wings. It was this sort of army that fought the Etruscans and Samnites during the 5th and 4th centuries B.C. It is said that these foes used the pilum, a seven-foot javelin which the Romans, always ready to adopt anything that looked good, took up and turned to the fearsome use that was probably more responsible for the success of the Roman legionary than anything else. Experience was teaching the Romans the advantages of a more open order than the phalanx and the adoption of the pilum forced them to break up the old solid formations into smaller units as it was difficult to throw the javelin from closely packed ranks.

It was probably in about the 4th century B.C. (at about the same time that the Macedonians were developing their phalanx of sarissae) that the Roman legionary formation was adopted. The structure of the legion was entirely recast so that it was divided into three sections—the hastati (young soldiers); the principes (men of mature age) and the triarii (the veterans). Each section was sub-divided into ten maniples formed of 120 men each in the first two cases whilst the triarii was 60 strong. Attached to these heavy infantry were 1,200 lightly armed troops (velites) and 300 cavalry. The normal strength of the legion was 4,500 men, sometimes enlarged to 6,000 men, and usually four legions were raised for a campaign, two commanded by each consul assisted by a staff of military tribunes, six to each legion. Each maniple was officered by two centurions.

The Roman cavalry of those days was largely composed of members of the aristocracy and was formed into a tactical unit, the turma, made up of three ducuriae of 10 men each. There were

10 turmae to each legion, usually positioned on the wings but sometimes held back in reserve.

In battle, the legion formed up on a frontage of just over 300 yards with maniples arranged in a chess-board formation with their men formed 12 wide by 10 deep. Each maniple was an independent unit which could be manoeuvred separately and if a gap appeared in the front line a maniple could be pushed forward to fill it, the chess-board formation making it easy for rear divisions to pass through to the front. The legion was a flexible phalanx, combining solidity and divisibility, pliancy and strength, weight and manoeuvring power so that it did not break up when penetrated by the enemy, or if part of it was overwhelmed then the remaining maniples closed up and carried on the fight.

The Roman method of fighting was first to send forward the light infantry in skirmishing order, then the hastati charged, hurling their javelins and attacking the enemy with their swords whilst the principes and triarii followed up in support.

Roman chequered formation of the 4th century

The arms and equipment of the legions did not vary to any great degree throughout their centuries of supremacy. The sculptured pictures on Trajan's Column in Rome depict a soldier wearing a round helmet with cheek guards, shoulders protected by five hinged iron strips that allowed free movement to the arms whilst from waist to upper chest the body was guarded by similar strip-armour, jointed so as to allow the soldier to move freely when fighting. Under his armour he wore a kilt-length shirt with sleeves reaching down to the elbows and under the kilt he wore a pair of tight woollen breeches (sometimes these are said to have been made of leather). On his feet he wore thonged marching sandals. Officers of centurion rank and above were dressed in a far more elaborate fashion with plumed helmets and beautifully carved or moulded breast-plates—probably provided at their own expense. It is said that Julius Caesar approved of this because it both improved their appearance and made them careful in battle. The legionary carried a large rectangular shield made of wood covered with leather and bent into the shape of a half cylinder.

He was armed with two pila, both of which played a most

significant part in the success of the Roman soldier. Already mentioned, the pilum was a heavy five- to seven-foot-long throwing spear with a sharp tempered point connected to the wooden shaft by a long shank of soft iron that played a dual rule. Once the spear had been thrown (because of its weight it could only be sent perhaps 20 paces) it would either cause a grave wound or would be deflected by the target-warrior on his shield. But once embedded in the shield, the weight of the shaft bent the soft-iron shank so that it dragged down the shield leaving its bearer either defenceless or saddled with an unmanageable and clumsy shield. Also, the bending of the weapon rendered it useless for throwing back at its original owner. Probably each soldier carried two pila. The second weapon used by the legionary was a short sword (gladius) with a blade about 18 to 20 inches long, two-edged and pointed so that it was a thrusting weapon. It was particularly deadly at close quarters when the soldier held the hilt well down and brought the point smartly up at the enemy's head or face. As a thrust is far more likely to be fatal than a slashing cut, the gladius in the hands of the highly trained legionaries was a close-quarters weapon almost without parallel. It was with these weapons that the Romans fought against Carthage in the Punic Wars.

The contrasting methods of warfare of the Greeks and the Romans were demonstrated when Pyrrhus, King of Epirus, confronted them with his armies at Heraclea in 280 B.C. and Ausculum in 279 B.C. At Heraclea, after driving the Greek cavalry off the field, legion and phalanx are said to have clashed seven times without decisive result, but just as the battle was turning in favour of the Romans, Pyrrhus brought up 20 war elephants. Never having seen such monstrous animals before, the Romans were thrown into confusion so that their fleeing cavalry pursued by the elephants broke the ranks of the legionaries and enabled the Greeks to gain the upper hand. However, the losses were so great on both sides that it gave birth to the term 'a Pyrrhic victory'. Although Pyrrhus again defeated the Romans in 279 he was heavily defeated at Beneventum in 275 B.C. when the Romans actually captured four of his war elephants. Later wars waged by the Romans against the successors of Alexander the Great in Greece and Asia saw the legion gain an increasing supremacy over the phalanx. But it was hardly a fair trial because the phalanx was no longer at its best and opposed to the mobile and flexible legion it was like a bull surrounded by fleet-footed toreadors. Alexander had used his cavalry to protect the flanks of his phalanx but his successors ignored this vital fact and rested their confidence in the known fact that against a foe who waited to be attacked on level ground the phalanx was invincible. But at Cynoscephalae in 197 B.C. the Romans lured the phalanx on to rough ground so that it became disordered and was taken in rear by 20 Roman maniples who had marched across the battlefield. At Magnesia in 189 B.C. Roman cavalry hit the phalanx in the flank and destroyed it and at Pydna in 168 B.C. the phalanx advanced too hastily, leaving a gap in their centre into which the Romans charged. This was probably the last battle in which the phalanx played any serious part in warfare.

After the Punic Wars the city-state of Rome had become a world power, changing from a Republic to an Empire. Inevitably this revolutionised the Roman army because of the need for expeditionary forces serving long periods overseas. From 117 B.C. onwards the Romans were threatened for several years by two barbarian nations, the Teutones and the Cimbri—savages like the Sudanese warriors of the 19th century who charged with stark courage and great physical strength that was almost irresistible. Said to total 300,000 they overran Gaul and defeated army after army of Romans until the most crushing blow of all at Arausio in 105 B.C.; they destroyed two consular armies of 80,000 men so that only 10 men escaped. Comparable to the disaster at Cannae, this defeat showed the need for changes in the military methods of the Romans. Fortunately they were given about three years of grace in which to make these changes whilst the barbarians fought the tribesmen of northern Spain.

Marius, a great Roman general (157 to 86 B.C.), made full use of the period to professionalise the Roman army; he eliminated the property qualification and enrolled all suitable volunteers for long periods of service of up to twenty years. Marius did away with the three divisions of heavy infantry (the hastati, the principes and triarii) and replaced the loose lines of maniple with a denser formation, a tactical unit with a broader front and narrower intervals, which he called a cohort. Each cohort consisted of 600 men divided into six centuries with ten cohorts to a legion. From now on the Marian or cohort legion became the standard fighting unit of the Roman army. At this time legions began to be numbered, a practice which persisted until the fall of the Empire.

Another Marian innovation was the withdrawal of Roman light troops and cavalry from the legion—for these and other supernumeraries Rome now relied on the provincial auxiliaries. Among them were Gallic and African light infantry, Cretan archers, slingers from the Balearic Islands, whilst the cavalry were mostly Gauls or Germans. Attached to each legion were 300 cavalry either divided into ten squadrons, armed with a shortish heavy spear plus eight javelins and a long sword (the spatha). These auxiliaries were to be a regular feature of the Roman army for centuries to come, fighting in their own native manner, using their own weapons with all the dash of the barbarian. It was this professional allegiance that conquered Gaul under Caesar and fought the Civil Wars that ended with the establishment of the Empire. Well disciplined, confident to the point of arrogance, this mobile legion was a magnificent fighting machine. Not only were its cavalry and infantry tactics of the highest order but it was adept at siege work and fortifications—its temporary and permanent camps were heavily fortified and laid out in a regular,

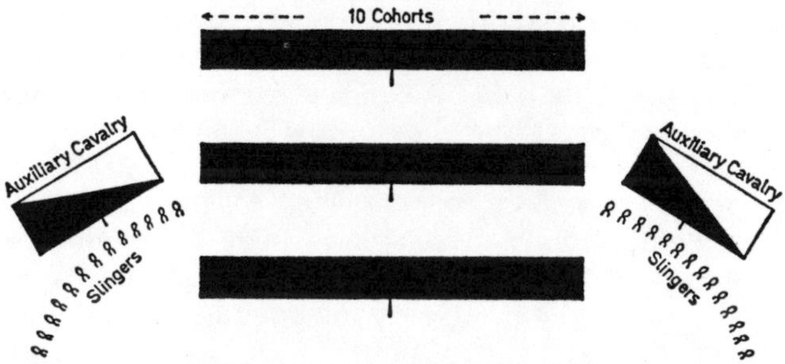


The Roman legion battle formation

rectilinear manner. The Roman soldier himself performed the manual labour of fortifying camps and building bridges under the direction of specialist engineers attached to each legion. It is believed that the artillery used for defence of fortified positions and in sieges was manned from the ranks of the legions. This artillery included battering rams, movable towers, missile-firing artillery and catapults. Many of these instruments of war were built on the spot where they were required to fight from prefabricated iron parts and wooden frames carried with the baggage.

Like everything else they did, the Roman method of fighting was a laid down procedure, tested and tempered on many battlefields against all varieties of foe. When battle seemed imminent the ten cohorts formed up three deep on a frontage of 2,000 feet which allowed each individual legionary six feet of space in which to fight. On the flanks and to their front the auxiliary slingers and archers skirmished and kept up hails of stones and arrows on the enemy. Out on the flank the auxiliary horsemen drove in on either side with the idea of bunching up the enemy for the coming attack. It was a cardinal rule that when the enemy was in a defensive position the Roman attack began when 360 feet away or 750 feet away if the enemy was advancing—it had been worked out that in either case the legionary had to cover the same distance. The first front line of cohorts moved forward at the march and then at the double until the first two ranks of each cohort were within about 60 yards of the enemy when they would launch themselves forward and hurl their pila at a range of about 20 paces. At the same time their comrades followed up and repeated the performance until the back lines of the cohorts had thrown their pila over the heads of their front-rank comrades who were now among the enemy thrusting murderously with their short swords. Whilst this was going on, the second line of cohorts were moving forward their front men preparing for their 60-yard run. The third line of cohorts were probably held back in reserve to be used to stem the tide if the battle went against the Romans or to mop up the enemy's last efforts at resistance.

The legion was not at its best when relentlessly attacked by hordes of highly mobile cavalry or infantry discharging javelins and arrows at them. With the heavy pilum outranged by arrows and lighter javelins, the relatively ponderous advances of the legion were countered by the enemy's faster withdrawals. At Carrhae in 54 B.C., the Parthians lured a Roman army into the desert and destroyed it by never allowing its light horse archers to come to grips but riding in, shooting and retiring when threatened. The Parthians were largely a mounted army consisting of heavily armed cavalry wearing armour and carrying long lances and powerful bows. Their more numerous light cavalry were unarmoured but were highly adept at firing arrows from their moving horses. The Romans formed up in defensive positions and sent forward their auxiliaries who were quickly forced back by showers of arrows. Then followed a nightmare period as the Roman ranks stood firm under a piltiless rain of arrows poured into them by hordes of circling archers. A desperate charge by the Romans auxiliary cavalry was blocked off by the heavy Parthian cavalry and then shot to pieces by archers in a desperate retreat; the Roman lost about 20,000 men killed with 10,000 taken prisoner.

An attacking not a defensive weapon, the legion fought their defensive actions in single-line square or circle with their light troops probably in the centre. Handicapped by their lack of missile power, the Romans placed great trust in fortified earthworks when defensive tactics were necessary.

On level ground and against undisciplined tribesmen, the legion was almost irresistible where their fighting methods made it possible for fresh troops to continue the battle and keep the enemy at full stretch for long periods. On the assumption that a legionary in full battle gear could only fight at peak efficiency for fifteen minutes, a cohort of 600 men was divided into ten 60-man waves each fighting for fifteen minutes and then retiring to rest until their turn to fight came round again.

Although the legions of Augustus differed little in equipment and tactics from those of Caesar, the character of the army itself had changed. Gradually it was becoming an army of defence dotted along the vast and ever expanding frontiers of the Roman world. This meant there was seldom any groupings as large as a field army, rather it was a question of a legion being stationed in an area or a cohort of auxiliary troops in an important town. By the 2nd century A.D. the auxiliaries no longer retained their native weapons but were trained and armed in Roman fashion although their arms and equipment varied from that of the regular legion in that they carried a lighter spear than the pilum and the long broad-sword, the spatha.

Nevertheless, for many years the legions posted around the Empire, from Scotland to the Euphrates, managed to hold back the barbarian hordes. From the 2nd century A.D. onwards, when the era of conquest was definitely over, this meant that the legion tended more and more to adopt defensive tactics. Following a policy which gradually sapped their vigour and relaxed their discipline—a deliberate abandonment of offensive strategy and tactics always reacts unfavourably on the fighting quality of an army. The almost constant fighting was of a typical frontier nature, consisting of raids and punitive expeditions and the quelling of tribal uprisings and local rebellions. Always fewer in number than the barbarians they attempted to hold back, the legions were forced in to a defensive posture, hand-to-hand fighting became rare, the sword was discarded for the spear and missile weapons were increasingly used. Frontiers were planned so as to take the fullest advantage of natural obstacles such as hills and rivers and artificial obstacles such as earthworks, forts, block houses and even walls like Hadrian's Wall helped to hold back the invaders.

In spite of temporary lapses and occasional defeats, the legions were able to beat all weapons used both separately and in combination by infantry, horse cavalry and even elephants until late in the 4th century. With the increasing pressure of the barbarian nations the ancient frontiers began to shorten from the middle of the 3rd century. Whole nations were on the march and no frontier post or walls thinly held by legionaries and auxiliaries could hope to hold the masses flung against weak links in the chain. Field armies had to be formed and the bottom of the barrel was scraped to make up their numbers. The legions were depreciating not only in size but in quality—the harsh discipline, the long service and slow promotion together with the heavy arms and equipment made service as an auxiliary far more attractive and the barbarisation of the legions (by allowing barbarian tribes to serve under Roman standards) destroyed their discipline and efficiency. The barbarians who entered the Roman service were mainly mounted warriors and the imposition of their fighting methods on the Roman generals (a process helped by the steady deterioration of the legionary infantry) together with the need for mobility to combat the faster moving barbarians caused the trend towards large numbers of cavalry. Towards the end of the 4th century A.D. the Gothic cavalry had heavily defeated the Romans and many were persuaded that from now on Rome had to adopt heavy cavalry or perish. The infantry equipment grew lighter (by the 4th century the heavy pilum was no longer carried and had been replaced by the pike) and some of the cavalry (the cataphracti) had both man and horse covered with armour.

This replacement of infantry by cavalry was a logical change although contra to the tradition of centuries. The loss of discipline in the infantry made it necessary to adopt closer formations together with a weapon that would keep calvalry at a distance but massed infantry formations are not easily manoeuvrable and against an enemy who employed a large proportion of horsemen armoured infantry were too slow moving. Obviously the answer to this was more cavalry and in about 300 A.D. Diocletian raised a proportion of cavalry in the army from a tenth to a third. A hundred years later, infantry formed less than a half of the total numbers and were relegated to unimportant duties like scouting, skirmishing and garrisoning strongholds. The supremacy of the horse soldier, thus established, lasted unbroken for another thousand years. It was not long before the restless northern tribes of horsemen, the Goths, Vandals and Huns pushed southward to sweep the legions away, splitting Rome into a western Empire governed from Rome itself and an eastern Empire with its own Emperor in Constantinople.

## STYLE OF FIGHTING

### *Earliest Times*

Phalanx and maniple in chess-board formation, a tactical flexibility in the cellular battle order that was a marked deviation from the solid mass of the Greek phalanx. Originally adopted to permit easier movement in combat over uneven terrain, this flexible formation of well drilled men trained to manoeuvre reached its peak at the Battle of Ilipa (206 B.C.) when Scipio with 48,000 men decisively defeated 70,000 Carthaginians in a battle of brilliant manoeuvre.

'During the last half of the 3rd century B.C. the Roman Legion probably achieved its highest development and greatest competency' (Dupuy).

At about 50 B.C. the ever increasing commitments of the Roman army resulted in a lowering of the standards of recruits so reducing discipline and training with a subsequent lack of confidence between the commander and his troops. This resulted in a tendency to reduce the intervals between the maniples of the legion so that it began to resemble the old Greek phalanx in battle order. Obviously this detracted from the normal superiority of the Roman formation over those of its enemies and led to a number of Roman defeats. Possibly this was also to present a closer-knit formation against such as the German's wild charges.

### *The Marian Legion Period*

Cohort legion formation as detailed in text (see Dupuy, pages 96–102, for detailed description).

Major defensive formations of the legion were the line, square and circle. When formed behind fortifications or entrenchments the line was usually a single line of 10 cohorts. The square was formed from a normal three-line formation, with 3 cohorts facing front, 3 facing rear and 2 to each flank. Modifications of this formation and the circle were employed in defence against cavalry but if the legion's flanks were protected by friendly cavalry or light auxiliary troops, the legion preferred to face the cavalry in its normal line formation.

Whenever possible the Romans tried to obtain the important advantage of being on higher ground than the enemy so adding to the range of missiles, increasing the shock effect of a charge and reducing physical effect of making it and also making it more difficult for the enemy to attack. To get maximum effect from the initial shock of battle, Julius Caesar usually placed his best cohorts in the front line.

Exceptionally well drilled and trained, the legion was controlled with discipline and organisation that enabled it to manoeuvre smoothly on the battlefield and to keep up a constant replacement of the men actually in the fighting line. This was inestimable advantage when fighting barbarians and, together with good leadership, is the principal reason why small numbers of Romans consistently defeated much larger barbarian forces. It is unlikely that the Romans were as outnumbered in battles with barbarians as most accounts seem to suggest, perhaps because these accounts were written by Romans!

There were a number of 'Above-average' commanders leading the Roman armies at various times. In the Rules of the War Games Research Group, auxiliary troops under Roman officers count as Romans—it is left to the wargamer's discretion whether he counts his auxiliaries as 'Native' troops or as Romans. In the latter years of the Roman Empire, great numbers of mercenaries were employed—their conditions of fighting were different from that of the Roman legionary and should be allowed for.

See Dupuy, pages 144–50, for details of later Roman formations and tactics.

*Fighting Assessment Chart:* ROMANS

| *Factor* | *Rating* | *Points* |
|---|---|---|
| 1. Ability to assault in controlled order | First Class | 3 |
| 2. Bravery | First Class | 3 |
| 3. Control | First Class | 3 |
| 4. Defensive ability | First Class | 3 |
| 5. Desperation | Not applicable | – |
| 6. Discipline | First Class | 3 |
| 7. Ferocity | Average | 1 |
| 8. Manoeuvrability | First Class | 3 |
| 9. Mobility | First Class | 3 |
| 10. Movement, speed of | First Class | 3 |
| 11. Order | First Class | 3 |
| 12. Tactical Sense | First Class | 3 |
| 13. Training and Drilling | First Class | 3 |
| 14. United action, capability for | First Class | 3 |
| 15. Weapon Handling | First Class | 3 |
| 16. Demoralisation, likelihood of | Unlikely | 3 |
| 17. Impetuous Action, likelihood of | Unlikely | 3 |
| 18. Panic, likelihood of | Unlikely | 3 |
| 19. Reluctance to take definite action | Unlikely | 3 |
| 20. Unscheduled Manoeuvres, likelihood of | Unlikely | 3 |
| | *Fighting Assessment* | 55 |

# THE CARTHAGINIANS

The Punic Wars were a prolonged and bitter struggle between Rome's growing and ambitious military strength and Carthage, a North African trading state founded by the Phoenicians, and her powerful fleet. It is likely that the wars were brought about by the fact that the Mediterranean area was just not big enough for two great powers, aggravated by the Carthaginian claim of a monopoly in all Mediterranean waters and the Roman resentment of the Carthaginian presence in Sicily.

In the first Punic War (264–241 B.C.) neither side produced a commander of outstanding ability and the land fighting was indecisive. Nevertheless it proved that on land the Carthaginians were no match for the infantry of the legion. Carthage did not possess a citizen army as did Rome, but was protected by a collection of mercenaries, African subjects and allies engaged at short notice for a particular campaign. Under Carthaginian officers, these hard-fighting mercenaries included Libyans, Gauls,

Spaniards, slingers from the Balearic Isles, Greeks and Ligurians. The Carthaginian cavalry were equipped in the Greek style and included squadrons of light Numidian horsemen, riding without saddles and bridles and noted for fearlessness, endurance and cunning. Armed with the javelin, they did not close directly with the enemy but came within range and hurled their deadly missiles. Incongruously Rome, a non-maritime power, won the war at sea, wresting naval supremacy from the Carthaginians after realising that as long as Carthage controlled the sea she could reinforce and supply her troops in Sicily and also prevent any Roman invasion of her shores. In the beginning the Romans possessed no war ships that could stand up to the heavy quinqueremes of the Carthaginian navy. But with typical Roman doggedness and using a stranded Carthaginian quinquereme as a model, the Romans laboriously built themselves a fleet.

Attaining mastery of the seas did not come easy to the Romans and some of her early naval battles were a disaster. For instance, at the Battle of Drepanum in 249 B.C. Rome lost 180 out of her fleet of 210 ships and in the same year in other engagements lost another 120 battle ships and 800 transports. But the proud Romans would not be deterred and eight years later they launched a new fleet of 200 ships and beat the war-weary Carthaginians. They achieved this by revising a method of grappling and boarding the Carthaginian ships so neutralising their superior manoeuvrability and skilful seamanship that enabled them to ram or shear the oars off their opponents. The Roman method was to build ships as high as those of the Carthaginians and to equip them with a number of drawbridges wide enough for two men to cross abreast, hinged to the deck of the Roman ship and with a large spike on the underside of the outer end of the drawbridge. On contacting a Carthaginian vessel, the tackle holding the drawbridge upright was cut allowing the drawbridge to crash down on to the deck of the enemy vessel so that its spike pierced deeply into the deck timbers, so holding the vessels together whilst the heavily armed Roman legionaries surged across to overcome the Carthaginian sailors and soldiers by their superior armour and fighting ability. With control of the sea, the Romans were now able for the first time to send expeditions to land on enemy soil.

While winning the first Punic War the Romans secured possession of Sicily and later Sardinia and Corsica. Trying to recoup their losses, the Carthaginians founded an Empire in Spain but in doing so aroused the envy of Rome who found further pretexts to force them into a second war. The second Punic War (219–202

B.C.) is notable for the appearance of Hannibal, one of the most brilliant soldiers the world has known. Unable to invade Italy because of the Roman control of the sea, Hannibal staked everything on an invasion of Italy, leading his army from Spanish soil into Italy by the land route over the Alps. It was an incredible project that cost Hannibal 26,000 of the 46,000 men and 37 elephants that set out from Spain.

For the next three years Hannibal and his armies moved about Italy delivering a series of smashing blows to the Romans. It was a period when the Roman commanders were men of mediocre military talent so that their great superiority both in numbers and fighting power of the Roman infantry were more than off-set by the skilful manner in which Hannibal handled his cavalry. The Carthaginian infantry was relatively lightly armed with short swords, spears and shields and little defensive armour but their cavalry were far more numerous than the Roman horsemen. After beating the Romans at the Trebbia in 218 B.C., Hannibal marched south and ambushed and defeated a Roman army at Lake Trasimene in 217 B.C.

It was the Battle of Cannae (216 B.C.) where the Carthaginians numbering 50,000 almost completely annihilated the largest army that Rome had ever put into the field—some 76,000 men. Cannae has claims to be the high water mark of military achievement in the ancient world—its combination of originality and conception and artistry in execution makes it 'the perfect battle' where a strong force is surrounded and destroyed by a weaker one. With their right resting on a river, the Romans lined up with their cavalry on the wings and infantry in the centre, massed in deep narrow columns so that their depth exceeded their breadth. Hannibal had his heavy cavalry on the left and his light cavalry on the right (10,000 against the Roman's 6,000) and his infantry (40,000 against 70,000) in the centre. But Hannibal placed his infantry in a carefully designed position with his Gauls in the centre and his more heavily armoured Carthaginian infantry on their flanks in columns. The advancing legionaries pressed back the Gauls as Hannibal anticipated and entered the lane between the Carthaginian columns who immediately faced inwards on the side of the Roman flanks so that the legionaries found themselves surrounded on three sides by the enemy. In the meantime, the Carthaginian heavy cavalry scattered the Roman cavalry and then crossed the battlefield to aid the Numidian light cavalry in dispersing the remainder of the Roman horsemen. Then both turned and charged the legionaries in the rear, tumbling the maniples into disorder. Packed into a narrow space, lacking room to wield their weapons, their leaders gone and their formations shattered, the legionaries were slaughtered like sheep. They lost 48,000 dead against 6,700 of the Carthaginians, most of whom were Gauls.

With their armies destroyed and a seemingly unconquerable enemy firmly established on her soil it seemed as though the Romans were beaten. But the picture was not as gloomy as it might appear because Hannibal had failed to win over to his side the allies of Rome, nor was he strong enough to attack Rome itself and the Roman fleet prevented aid coming to him from Carthage. The Roman armies in Spain were more than holding their own, finally managing in 206 B.C. to drive the Carthaginians out. In 212 B.C. a revolt in Africa gave a considerable amount of trouble to the Carthaginians and in 214 B.C. they failed in an invasion of Sardinia and were unable to make much impression upon the Roman forces in Sicily.

Publius Scipio, the son of a Roman general killed in Spain, went to great lengths to remedy the obvious Roman cavalry weakness by equipping his cavalry with helmets, breast-plates and shields and arming them with lance, javelin and curved swords whilst placing great emphasis on manoeuvring at speed with unbroken ranks. At Ilipa on African soil in 206 B.C. he defeated a large Carthaginian army under Hasdrubal Gisgo by using his legionaries on the flanks, Spanish allies in the centre with cavalry and light troops sweeping round the Carthaginian flanks and attacking them in the rear. Disorganised by their own elephants who had stampeded through their ranks, the Carthaginians were so badly beaten that only 6,000 out of 70,000 survived.

Hannibal's brother Hasdrubal brought his army into Italy to reinforce Hannibal. Before they could affect a junction, Claudius Nero, the Roman commander, completely outwitted Hannibal by leaving a small force to occupy him whilst he marched rapidly north to join forces with another Roman army. Together they completely defeated the Carthaginians, killing Hasdrubal, and then Nero marched his army back 250 miles to form them again in front of Hannibal before the Carthaginian commander, knew of his brother's defeat at the Metaurus. This battle was the turning point in the Punic War and is listed by Creasy in his *Fifteen Decisive Battles of the World.*

Hannibal returned to Africa in 203 B.C. to come up against Scipio near Zama in 202 B.C. where the battle was fought that was the final act in the long sixteen-year-old drama. Each general commanded a force of about 46,000 men—Hannibal formed his infantry in three lines with his veterans of the Italian campaign in the third line as a reserve. His 80 elephants were distributed along his front. Smaller in number than the Romans, his cavalry were placed on the wings. Scipio placed his two veteran legions in the centre, his Roman cavalry on the left and his Numidians on the right. To combat the elephants he drew his maniples up one behind the other with wide lanes left through their formations. After some skirmishing with light troops, Hannibal sent his elephants forward to be met with blasts of trumpets which caused some to turn and run back through their own lines whilst others ran down the lanes between the maniples assailed by javelins and frightened by the shouts and screams of the blocks of legionaries. Taking advantage of the confusion, the Roman and Numidian cavalry charged the Carthaginian cavalry and chased them off the field. The Roman legionaries pressed back the Gauls and Ligurians who formed Hannibal's first line and then did the same to the Carthaginians in the second line. It has been claimed that this was what Hannibal had in mind, planning to throw his Italian veterans on the legions when they were disordered by their fierce struggles with the first two lines. Just in time Scipio realised Hannibal's intention and halted his men to re-form their ranks. Roman legion and Carthaginian phalanx crashed together, the former holding the superior numbers of the Carthaginians until their cavalry returned from the chase to fall upon the rear of the Carthaginian phalanx. Unable to resist these blows from front and rear, the Carthaginians crumbled and died fighting—Hannibal escaped from the field. With the last of her armies destroyed, Carthage accepted peace on Roman terms and the war was over.

This was not the end of the story because the Romans were determined to end once and for all the Carthaginian threat and for the next half century they did all within their power to provoke the Carthaginians. At last, disarmed and without allies, ships or weapons, the Carthaginians resolved to fight back. With improvised arms and equipment they gallantly defended their city for three years before the Romans broke in and captured the place street by street. Out of a population of half a million, 55,000 inhabitants survived and were sold into slavery—the city was completely destroyed in 146 B.C. and its whole area was ploughed over so that no trace of it remained.

## STYLE OF FIGHTING

Formed largely of mercenaries, professional soldiers who were well drilled, trained and disciplined, the Carthaginian army were a formidable force capable of manoeuvring under battle conditions.

Like the Macedonians and the Romans, the Carthaginian had learnt through experience to so organise their baggage trains as to eliminate all unnecessary equipment—they used pack animals, waggons and handcarts.

This period was notable for its high standards of military leadership. Hannibal's father Hamilcar was a very good general but the son was an inspired leader with exceptional tactical and strategic skills. Whilst he may not have been the greatest general of all time, it is impossible to rank the finest leaders of history very far ahead of him. Against the most military efficient nation in the world, he led armies formed of soldiers of lesser quality with a very considerable degree of success. The Roman Generals Scipio, Marcellus and Nero greatly added to their stature by learning from Hannibal.

Factors that must be considered when wargaming are (a) the use of mercenaries, (b) 'Above-average' commanders and, perhaps in the later stages, (c) a 'Desperation Factor' for the Carthaginians.

*Fighting Assessment Chart:* CARTHAGINIANS

| | *Factor* | *Rating* | *Points* |
|---|---|---|---|
| 1. | Ability to assault in controlled order | First Class | 3 |
| 2. | Bravery | First Class | 3 |
| 3. | Control | Reasonable | 2 |
| 4. | Defensive ability | Reasonable | 2 |
| 5. | Desperation | Average | 1 |
| 6. | Discipline | First Class | 3 |
| 7. | Ferocity | Not applicable | – |
| 8. | Manoeuvrability | First Class | 3 |
| 9. | Mobility | First Class | 3 |
| 10. | Movement, speed of | Reasonable | 2 |
| 11. | Order | First Class | 3 |
| 12. | Tactical Sense | Reasonable | 2 |
| 13. | Training and Drilling | First Class | 3 |
| 14. | United action, capability for | First Class | 3 |
| 15. | Weapon Handling | First Class | 3 |
| 16. | Demoralisation, likelihood of | Unlikely | 3 |
| 17. | Impetuous Action, likelihood of | Unlikely | 3 |
| 18. | Panic, likelihood of | Unlikely | 3 |
| 19. | Reluctance to take definite action | Unlikely | 3 |
| 20. | Unscheduled Manoeuvres, likelihood of | Unlikely | 3 |
| | | *Fighting Assessment* | 51 |

# THE ROMAN CIVIL WAR

## *50–44 B.C.*

The defeat and death of Crassus at Carrhae in 53 B.C. left Julius Caesar and Pompey as contenders for Rome and in the Civil War that followed the legions fought as well against the armies of Pompey as they had done against the barbarians. Crossing the Rubricon in 49 B.C., Caesar marched on Rome and gained a central position amid his numerically superior enemies who were scattered throughout Spain, Greece and Africa and, operating on interior lines, he struck at each of them in turn. The war spread to Spain, where the Pompeian refused battle until Caesar was placed in a desperate position through shortage of supplies, but a minor naval success so impressed the Spanish that they diverted their food supplies from the Pompeians to Caesar's camp. Eventually when they had been forced by hunger to retreat, Caesar moved rapidly across the Pompeians' rear and compelled them to surrender at Ilerda. Although the Pompeian navy controlled the Adriatic, Caesar slipped across to Thessaly with his army where he was cut off from Italy. Pompey counter-attacked and defeated the forces who were besieging him at Dyrrachium and then led his army into the interior after Caesar. But, at Pharsalus in 48 B.C., Caesar gained a tactical victory over them which released him from the trap. With the intention of overwhelming Caesar's cavalry and then rolling up his infantry, Pompey had massed 7,000 cavalry on his left flank against Caesar's 1,000 horsemen. To meet this blow, Caesar detached 6 cohorts from his infantry line and stationed them behind the right wing of the legions. As the Pompeian horse thundered past the infantry flank the cohorts suddenly hurled themselves against it in an unexpected attack that sent the enemy troopers galloping from the field. Supported by the remnants of Caesar's cavalry, the victorious cohorts moved forward and assailed the Pompeian infantry in flank and rear until, after a brief struggle, they broke and fled. By the most unorthodox military manoeuvre of infantry charging cavalry, Caesar had gained mastery of the Roman world.

### STYLE OF FIGHTING

At the Battle of Pharsalus (48 B.C.), outnumbered, Julius Caesar formed a line of battle with his left flank on a river, his legions in the customary three lines but with 6 cohorts covering his right rear, where the chief danger lay because his cavalry were outnumbered 7 to 1 by Pompey's horsemen. Caesar extended the intervals between the cohorts with his main body to match the frontage of Pompey's army which was drawn up in its normal formation. The third line of his main body was, as usual, kept in reserve to the first two lines. He then ordered his first two lines of infantry to attack the stationary army of Pompey. At the moment of impact, Pompey launched his cavalry supported by archers and slingers against Caesar's horse who fought stubbornly but were forced back by weight of numbers. At the decisive moment, Caesar personally led his selective force of 6 cohorts against the flank of Pompey's advancing horsemen and scattered the surprised cavalry, to push on, slaughter the archers and slingers and then turn against the left flank of Pompey's main body. Galloping to join his third line, Caesar led them through the intervals of the first two to smash into the front of the Pompeian legions—this charge, together with the surprise envelopment of the 6 cohorts, smashed the Pompeian army.

# THE GAULS

One of the first conquerors of Rome was Brennus the Gaul who led his warriors into Rome in 390 B.C. to sack it and humble its citizens. It was a defeat that remained unavenged for nearly 300 years until in 51 B.C., after seven years of fighting, Julius Caesar besieged and captured the town of Uxellodunum. To deter the Gauls from further resistance, Caesar made an example of the defenders of Uxellodunum and their commander Vercingetorix, after being kept in prison for six years, was led through the streets of Rome in triumph before being publicly strangled.

Like most of Rome's enemies, the Gauls found that bravery was not enough when confronted by military science. Nevertheless, they gave the Romans seven years of very hard fighting before being subdued. Gaul was a land covered with forests and marshes that was easy to defend and difficult to conquer. A hard warrior-people, the Gauls were inspired by the resolution and skill of Vercingetorix, so that it required never less than six legions, sometimes as many as eleven plus unknown numbers of auxiliary troops from other nations, to conquer the Gauls.

Caesar pursued a policy of utilising the jealousies and hatreds inevitable amongst the tribes and nations to play them off one against the other. This enabled him to raise local contingents led by their own chiefs and the Gauls furnished many of the auxiliary cavalry units that fought alongside the legions. It must have been confusing and it is said that Julius Caesar and his commanders often found it difficult, particularly during night attacks, to distinguish between their own men and those of Vercingetorix.

The Gauls were a disciplined and organised warrior race, well built and courageous men made strong by hunting and fighting. They fought in a loose phalanx, with the men standing close together, their tall narrow shields presenting a wall to the enemy whilst the ranks behind had their shields raised over their heads to ward off missiles. Such a closely packed and orderless formation was cohesive but it lacked mobility and meant that the bulk of the fighting fell on the men in the front who were pressed forward by the weight of those behind them. Their chiefs wore mail and many of the warriors had bronze or iron breast-plates and jerkins of scale armour. They wore iron or bronze helmets, often decorated with horns or spikes, bearing crossed feathers. The more primitive of them had no armour at all whilst others wore protective clothing and helmets made of leather. They were armed with light javelins that lacked the penetrating power of the pilum and a long two-edged slashing sword. They carried wooden or leather shields.

The strong arm of the Gauls was their cavalry who relied mainly on the shock charge delivered in a mass attack. They were equipped with javelins or lances and the long two-edged sword and often wore breast-plates of primitive plate, scale, or mail armour. They carried small round shields.

Perhaps Caesar was lucky in being allowed to gain experience of warfare against a barbarian enemy such as the Gauls who were rarely able to keep a large force together in the field nor to establish an efficient supply system. On the other hand, the Roman army possessed such a system and were always able to mass superior numbers at decisive points. Yet with all his superiority in numbers, armament and discipline, Caesar made some serious mistakes in Gaul and more than once narrowly escaped defeat. His pamphlet 'Gallic War' implies that he always fought against odds but modern research has established that the numerical strength of the Gauls as given by Caesar was invariably inflated and there is no doubt that he owed a great deal to the fighting qualities of his soldiers and their superior numbers.

### STYLE OF FIGHTING

In addition to attacking in a wild and orderless but extremely courageous assault, the Gauls had a modicum of discipline that enabled them to form a sort of 'shield-wall' and fight in a loose phalanx when on the defensive. This dense and irregular formation lacked mobility and could not manoeuvre under battle conditions.

*Fighting Assessment Chart*: GAULS

| *Factor* | *Rating* | *Points* |
|---|---|---|
| 1. Ability to assault in controlled order | None | 0 |
| 2. Bravery | First Class | 3 |
| 3. Control | Average | 1 |
| 4. Defensive ability (Very good behind fortifications, e.g. Siege of Alesia) | Average | 1 |
| 5. Desperation | Average | 1 |
| 6. Discipline | Reasonable | 2 |
| 7. Ferocity | Average | 1 |
| 8. Manoeuvrability | None | 0 |
| 9. Mobility | None | 0 |
| 10. Movement, speed of | Reasonable | 2 |
| 11. Order | Average | 1 |
| 12. Tactical Sense | Average | 1 |
| 13. Training and Drilling | None | 0 |
| 14. United action, capability for | Average | 1 |
| 15. Weapon Handling | First Class | 3 |
| 16. Demoralisation, likelihood of | Average | 1 |
| 17. Impetuous Action, likelihood of | Likely | 0 |
| 18. Panic, likelihood of | Average | 1 |
| 19. Reluctance to take definite action | Average | 1 |
| 20. Unscheduled Manoeuvres, likelihood of | Likely | 0 |
| | *Fighting Assessment* | 20 |

# THE GERMANS

For many years western tribesmen were a menace to the safety of Rome itself and prominent amongst these savage warriors were the formidable Germanic races from the immense forests that stretched from the Rhine to the Vistula, from the Baltic Sea to Bohemia. These early Germans were tall powerful people with long flaxen hair and bright blue eyes. They were all soldiers and their villages were grouped together to form a 'hundred'. The men of each hundred formed a battlegroup of 5 or 6,000 led by the chief of the hundred. They were a warlike people of fine physique hardened by their way of life with its emphasis on hunting and warfare.

The German cavalry were rated even higher than the Gallic horsemen and when the wars with Vercingetorix cut off Caesar's supply of Gallic cavalry he turned to the German tribes which he had pacified and raised a considerable force of horsemen. They were armed with a short spear, sometimes used as a lance and sometimes thrown like a javelin and, like the Gauls, they used the long two-edged slashing sword. Riding bare back without even a saddle pad or blanket, they had little defensive armour and many wore only skins and carried shields of plaited wicker-work. Renowned for their size and ferocity, these horsemen often jumped from their horses and fought on foot, training their mounts to stand and wait for their rider's return.

Ariovistus, a noted German leader, led a force of 6,000 horsemen and the same number of infantry about whom Caesar wrote:

The latter, particularly good runners, and remarkable for their courage, had been chosen out of the whole army, each cavalryman picking his own infantryman to look after him. When the cavalry were in difficulty they fell back on this body of infantry and the infantry would quickly concentrate. The foot soldiers would rally round any horseman who was wounded and had fallen from his mount. Long training had made them so fast on their feet that in a prolonged advance or a quick retreat they could keep pace by running alongside the horses and clinging to their manes.

The German tactics were those of mass assault, a technique of swamping the opponent regardless of loss in a lightning charge aimed at one corner of the enemy line and with the maximum strength to be deployed in a sudden onslaught. If the corner of the line gave way under the assault and a breach opened up, this was the beginning of victory. In a frenzy, shouting and chanting, wave after wave of wild horsemen and fleet-footed infantry, wielding maces and broad axes, burst upon the enemy. Caesar has written of it:

When the signal was given, our men rushed forward so fiercely and the enemy came on so swiftly and furiously that there was no time for hurling our javelins. They were thrown aside, and the fighting was with swords at close quarters. The Germans quickly adopted their usual close formation to defend themselves from the sword thrust, but many of our men were brave enough to leap right on top of the wall of shields, tear the shields from their hands that held them, and stab down the enemy from above.

Often during the battle, the women together with the old people and children hid behind the stockade made of carts and chariots, keeping up their fighters' spirits by beating furiously on gongs and drums made of sheets of leather.

It is worth pointing out that the Romans never considered the barbarians' brave and spectacular cavalry attacks to be capable of breaking formations of steady legionaries presenting an unbroken front—infantry caught in disorder or with an unprotected flank stood little chance however.

The course of history was changed when an army of barbarians under Arminus defeated and cut to pieces a regular Roman army of three legions in the Teutoberg Forest in 9 A.D. Tricked into making a lengthy march during the wet season through difficult marshy country, the Romans were caught in the dark ravines and defeated so decisively that their numbers never again appeared on the Roman army lists. It was a disaster that ended for all time Roman plans for including Germany in her great Confederacy. The Romans never completely subjugated the Germans but managed to maintain a considerable influence amongst them for a number of years.

## STYLE OF FIGHTING

Until the time of Caesar even veteran Roman legionaries had an unreasoning fear of Gauls and Germans, perhaps because these barbarians were so physically superior to the small and 'chunky' Roman soldier.

The German style of fighting consisted of a mass assault, planned and to a certain extent controlled because it was aimed at a corner of the enemy line of the 'hinge' between two formations. Regardless of loss, the German infantry and horsemen flung themselves forward in a succession of waves. German tribesmen showed an uncontrolled fury and bravery but it was not enough to compete successfully with Roman skilled discipline and organisation. They were not above trying feigned retreats.

*Fighting Assessment Chart:* GERMANS

| | *Factor* | *Rating* | *Points* |
|---|---|---|---|
| 1. | Ability to assault in controlled order | Average | 1 |
| 2. | Bravery | First Class | 3 |
| 3. | Control | Average | 1 |
| 4. | Defensive ability | None | 0 |
| 5. | Desperation | Average | 1 |
| 6. | Discipline | Average | 1 |
| 7. | Ferocity | Average | 1 |
| 8. | Manoeuvrability | Reasonable | 2 |
| 9. | Mobility | Average | 1 |
| 10. | Movement, speed of | Reasonable | 2 |
| 11. | Order | Average | 1 |
| 12. | Tactical Sense | Average | 1 |
| 13. | Training and Drilling | None | 0 |
| 14. | United action, capability for | Average | 1 |
| 15. | Weapon Handling | First Class | 3 |
| 16. | Demoralisation, likelihood of | Average | 1 |
| 17. | Impetuous Action, likelihood of | Likely | 0 |
| 18. | Panic, likelihood of | Average | 1 |
| 19. | Reluctance to take definite action | Rare | 2 |
| 20. | Unscheduled Manoeuvres, likelihood of | Average | 1 |
| | | *Fighting Assessment* | 24 |

# THE GOTHS

Originally from Sweden, the Goths were blond barbarians who, in about 200 A.D. had moved south in chariots with their wives and children. In the 4th century A.D., although they were known to use skilled foot archers, it was their cavalry who have made their mark in history as they altered the art of war by increasing the power of attacking cavalry. Protected by armour made out of iron plates or thick leather, the Goths used the spatha and are also reputed to have been skilful with the lasso. It was the attacks carried out by their closely packed squadrons of lancers and swordsmen that destroyed the Roman army of the East, killing the Emperor Valens, all his generals and 40,000 of his men at Adrianople 150 miles north-west of Constantinople in 378 A.D.

Often regarded as the beginning of the cavalry age, this Gothic victory brought to the Romans the sad realisation that the day of the Legion was finished—from now on heavy cavalry had to be the main part of the army.

In the battle, the Romans were pressing home their traditional attack with the heavy infantry cohorts supported on either wing by small squadrons of horse against a small number of Gothic infantry behind a barricade of wagons. Suddenly and without warning a large force of Gothic cavalry, who had been away foraging, came thundering back to charge the Roman army in the flank. The Roman horse squadrons on the left wing, taken by surprise, were either cut down or fled, allowing the Gothic cavalry to crash into the legionary formation, rolling the left flank in on to the centre so that they formed a confused, madly-fighting mass. The Gothic infantry, coming out from their laager, attacked the Roman infantry in front. At this, the Roman cavalry on the right wing wheeled away and fled from the field, leaving the legionaries to their fate. Assailed on all sides, unable to extend their ranks and too crowded to use their weapons, 40,000 Romans were butchered along with their Emperor.

Together with the Vandals, the Huns and other restless northern tribes, the Goths pushed southwards on their horses to sweep the legions of Rome away before them. But before that there had been a gradual infiltration of barbarians into legions themselves and by the late 4th century the cavalry arm of the Roman army was made up almost entirely of Goths.

## STYLE OF FIGHTING

The emergence of the armoured lancer, riding a heavy horse, revolutionised warfare as infantry formations found themselves unable to withstand the terror and shock of a heavy cavalry charge. Disciplined to a reasonable degree, it is conceivable that the Gothic cavalry might occasionally have been subject to the same pell-mell tactics as feudal cavalry.

When on campaigns, the Goths moved in great wagon convoys, assembling them each night in circular laagers. These were useful for defence and also provided ready-made bases for foraging and plundering. On occasions, the Goths strung their wagons together while on the march to create moving forts.

*Fighting Assessment Chart:* GOTHS

| Factor | Rating | Points |
|---|---|---|
| 1. Ability to assault in controlled order | Reasonable | 2 |
| 2. Bravery | First Class | 3 |
| 3. Control | Average | 1 |
| 4. Defensive ability | Reasonable | 2 |
| 5. Desperation | None | 0 |
| 6. Discipline | Reasonable | 2 |
| 7. Ferocity | Average | 1 |
| 8. Manoeuvrability | Reasonable | 2 |
| 9. Mobility | Reasonable | 2 |
| 10. Movement, speed of | Reasonable | 2 |
| 11. Order | Reasonable | 2 |
| 12. Tactical Sense | Reasonable | 2 |
| 13. Training and Drilling | Average | 1 |
| 14. United action, capability for | Reasonable | 2 |
| 15. Weapon Handling | First Class | 3 |
| 16. Demoralisation, likelihood of | Average | 1 |
| 17. Impetuous Action, likelihood of | Likely | 0 |
| 18. Panic, likelihood of | Average | 1 |
| 19. Reluctance to take definite action | Average | 1 |
| 20. Unscheduled Manoeuvres, likelihood of | Likely | 0 |
| | *Fighting Assessment* | 30 |

# THE BRITONS

In 55 B.C., Julius Caesar, with a force of two legions plus auxiliary cavalry, crossed the straits of Dover and invaded Britain, pushing only ten miles inland before rough weather damaged his ships so forcing him to retreat. Next year he returned with five legions and drove across the Thames to penetrate into what is now Hertfordshire. Caesar wrote about his battles with the British, dwelling at some length on the chariots which came to meet him the moment he landed.

The Britons begin by driving their chariots all over the field hurling javelins . . . the horses and the noise . . . are sufficient to throw their opponents ranks into disorder. Then, after making their way between the squadrons of their own cavalry, they jump down from the chariot and engage on foot. In the meantime their charioteers retire a short distance from the battle so that they are handy in case of retreat. Thus they combine the mobility of cavalry with the staying power of infantry.

The chariot used by the British was open in front with two spirited native ponies harnessed to a broad, flat pole. Two fighters rode in the chariot including the noble who drove and commanded it—sometimes one of them would run out along the pole to fling javelins from between the horses' heads. In action, the tactics were to race the chariots in a column straight at the opposing foot soldiers, sending a shower of short javelins amongst them if they broke. If they did not break, the chariots would wheel off along the front with the fighters flinging their javelins as they passed —sometimes they would circle the enemy formation throwing in another volley of javelins at their rear. After several attacks of this kind, the chariot fighter dismounted and went in with the infantry. Caesar was impressed by the way British cavalry and chariots retired into the woods when they were repulsed, and by 54 B.C., when he returned to Britain, Caesar had worked out the correct tactics to use against British chariots. He decided to attack them on sight before they could draw up into a defensive formation. Throughout this campaign Cassivellaunus, the British king, constantly harried the Roman columns with his chariots so that the legionary columns had to be protected by outlying cavalry. Much as they harassed the Roman auxiliary cavalry and Caesar's foraging parties, time proved that chariots could not break the solid ranks of legionary veterans.

The story of Julius Caesar's campaigns in Gaul indicates many alliances of tribes formed against him. Although many of the British tribes were akin to the Gauls they showed a complete inability or disinclination to unite against a common foe, except for a brief union of the Atrebates, Cantii, Catuvellauni and Trinovantes at the time of Aulus Plautius's invasion. The Romans made a point of exploiting this failing and their governors were strictly enjoined to avoid any military or political action which might encourage the tribes to unite.

After the death of Julius Caesar the Britons were left alone for almost a century until in 43 A.D. Claudius came with an invading force. Claudius believed the British warrior, man for man, to be at least equal to the Roman but considered that his methods of fighting and his arms were more suited to single combat. The Briton fought with a broad-sword, an unwieldly weapon at close quarters, and his leather buckler was too small to provide much protection against the powerful pilum. It was during this campaign in September 43 A.D. that the Romans won an extraordinary victory in the misty marshes near Romford in Essex. When the first column of British chariots rattled down upon the enemy their horses were thrown into confusion by a foul smell that filled the air—it came from a line of camels, creatures completely unknown to the Britons. With the leading chariots milling round in confusion, their horses practically out of control, the legionaries showered them with pila. Other chariot attacks were thrown back by showers of blazing balls of pitch thrown from war engines together with hails of missiles from the auxiliary slingers. Trip ropes had been placed in position during the night and the horses and chariots piled up on them. To add to the confusion the Britons found themselves attacked by coloured auxiliaries whose unfamiliarity caused them to be treated as black devils. It is said that, on at least one other occasion, the Romans brought elephants to Britain, with no doubt immense moral effect. Claudius used them for effect but never in action against the Britons.

The main military highways of the Roman province of Britain all radiated from London. The North Way ended at Lincoln; the Midland Way reached Wroxeter by way of Verulam and the great forests of central Britain; the West Way ran to Gloucester through Silchester. The main lateral road, the Frontier Way, ran from Exeter through Cirencester and Leicester to Lincoln. Its title was a misnomer because at the time of the rebellion of 61 A.D. the frontier had advanced to the line of a road adjoining Gloucester and Wroxeter, bordering the countries of the Silures and Ordovices. Other highways connected important towns—Colchester to Verulam; Silchester to Bath and Dorchester to Winchester and Portus. Planned and constructed under the supervision of military engineers, more roads were constantly being built by labour provided from tribes through whose territories they passed and they were also responsible for their subsequent maintenance.

It was over these roads that the Roman province of Britain maintained its communications by using hard-riding horsemen on fast, high quality horses working to a daily schedule of 50 miles. By means of rapid changes of mount, it was possible to send messages as fast as 80 miles a day.

In 61 A.D., the Roman Province of Britain was governed by Suetonius Paulinus, whose forces were disposed as follows:

At Wroxeter was the most powerful fortress and largest settlement in Britain, which dominated the upper Severn from a small plain encircled by foot hills that rose westwards to the mist-shrouded peaks of Ordovicia, half a day's march distance. This powerful fortress with a rampart two miles in perimeter housed two legions, XIV Gemina and XX Valeria, together with four auxiliary cavalry regiments and six auxiliary cohorts of infantry—a total of 17,000 men.

At Gloucester were II Legion Augusta plus 3,000 auxiliaries formed in three cavalry regiments and five infantry cohorts, one of them milliary. Detachments from all these units amounting to nearly 1,000 men were scattered at road stations along many of the highways of Britain.

At Lincoln was the Legion IX Hispana plus three auxiliary cavalry regiments and four auxiliary infantry cohorts. Total strength, legionaries and auxiliaries, was just over 9,000 men.

The auxiliary strength of the Roman army in Britain was 17,000 men, compared with 22,000 legionaries. At the time of the 61 A.D. rebellion, the Romans were considering recruiting from the Dobuni and Cornovii, two aboriginal tribes, who having been dispossessed of valuable territories by the invading Belgae tribe to the east and harried for generations by mountain tribes from the west, were the natural enemies of every other Briton in the province.

It was considered, at this time, that no tribes on the western or northern frontiers, other than the Brigantians and the Silures, were likely or expected to give trouble to the Romans. The former were a strong, well-armed tribe of fierce fighters who could put between 20/30,000 armed men in the field. The Brigantes were tall powerful men with fair skin and red hair, who wore only rough woollen trousers, bound criss-cross below the knee, otherwise they were naked. They were armed with either a long pointless sword used only for slashing, or a spear and a square wooden shield. Their chieftains wore brightly enamelled helmets and shields and brilliantly coloured cloaks. Their sword hilts, scabbards and saddlery were intricately designed with gold and silver and they wore jewelled brooches fastening their cloaks and belts. Some Brigantians fought from garishly painted two-horse chariots and they had a force of spearmen who rode bareback on small rough-coated ponies which they handled expertly. The Brigantes tribe lived in the area around what is now York.

The Silures tribe occupied the area north of the Severn, now South Wales. They were an older and different race to the others in the province, perhaps they were the survivors of the dark people who had built Stonehenge and had then been suddenly driven to the western mountains, where they now lived, by successive waves of invaders. Almost alone of all the tribes in Britain, they were consistently treacherous, lacked any understanding of diplomacy and disregarded treaties when it suited them, readily accepting gold and then immediately renewing hostilities with redoubled vigour. Like most other tribes they were divided into septs and clans each ruled by a minor chieftain from a hilltop stronghold but their king was worshipped like a God whose supreme power came from hereditary backgrounds.

Forming the backbone of Caractacus's army when he made his last stand against Scapula, the Silures were utterly defeated and, believing them to be in complete submission, the Romans were horrified when, completely without warning, they burst out again to attack and destroy detachments and burn the forts that were being built at strategic points over the countryside. Hurrying forward in support, the Legion's main body were completely defeated with the loss of the camp prefect, eight centurians and 700 men, the survivors being besieged in Gloucester until the tribesmen eventually dispersed to their homes. From then on the Severn was considered to be the limit of Roman dominated territory in this area. Convinced of their superiority in battle, the Silures incessantly harassed the garrison by attacking foraging parties and then ambushing rescuing parties of cavalry or auxiliary infantry. By escaping into the hills the tribesmen usually evaded the heavy legionary infantry sent out to punish them. As time went on their activities became more numerous and extensive, causing the area on the Roman side of the Severn to become denuded of people and their raids even threatened Cirencester. During these excursions they carried off prisoners, cattle and plunder, destroyed buildings and everything that they could not carry away and rarely were intercepted on their return journey. Other tribes were effected by their successes, the Dobuni, who were the chief sufferers from the Silures raids, eventually despaired of any protection from the Romans and attempted to make separate terms with the Silures. Didius Gallus brought twenty XX Legion from Wroxeter to support Augusta and set up a regular campaign to clear Dobunian territorial raiding parties. He raised forts at close intervals on the edge of the escarpment overlooking the Severn so that early warnings of raids were given and the expeditions could be sent out to cut off the Silures on their return journey. This policy began to achieve success and many Silurian war parties were completely destroyed so that not a man would return across the river. Nevertheless the tribesmen still remained aggressive and occasionally raided across the river although by now it was an unprofitable business because they had already destroyed all the settlements that had existed. In 61 A.D. all the land west of the Severn was Silurian and Roman forces could only enter their territory at the risk of almost immediate attack. Roman expeditions into Silurian territory were desperate affairs—the tribesmen always appeared to have advance warning of the attack and, abandoning their villages, garrisoned their numerous hill-forts. The Romans burnt the wooden villages but storming the forts was both laborious and expensive in manpower because every hill crest seemed to be crowned by a fort so that there was always another one to be taken. The only truly effective way of destroying the tribesmen was to catch them out in the open in the valleys but the Romans had never managed to do this. The Silures did not take part in Boadicea's rebellion because the Romans held hostages from their tribe.

In a bulge above the Thames Estuary, north of the Roman colony at Colchester, was the Icenian kingdom, which had Norwich as its capital. It was practically covered with marsh or forest and its people still lived as small communities in and around hill-forts. They were an independent, belligerent tribe who, like most British tribes, regarded their chieftain almost as a God. They had their own ruler, ran their own affairs and there were no Roman military posts in their territory. The Romans were conscious that Icenia lay on the flanks of the North Way, one of the main Roman lines of communication from London to Lincoln, and that the majority of the Roman army was stationed in the west so that the Iceni, a large and self-governing tribe, were far in the Roman rear. It was the intention of Suetonius Paulinus that the Iceni should be brought under Roman control. Chief of the Iceni was Prasutagus, an old and wise man who had already felt the strength of Rome's arm and was the only restraining influence that held the tribe back from attacking the Romans. After his death in 61 A.D., Boadicea, not without considerable provocation, led the tribe in revolt against the Romans.

As for the other tribes in Britain:

The Trinovantes, who lived north of Colchester bordering the Iceni, were one of the most civilised tribes in Britain but, like all citizens of a conquered and occupied country, were perhaps discontented by subjection and taxes.

The Degeangli tribe, who lived in the hills of northern Wales, were a small and usually peaceful tribe who had been easily defeated by a Roman expedition some ten years before the rebellion. At the same time the Ordovices tribe, who lived in Wales south of the Degeangli, had also been defeated by Roman expeditions. They were not a large tribe and if they were to give any trouble it would be because of the roughness of their terrain rather than their prowess in battle. But in 61 A.D. Suetonius Paulinus had led an expedition against the Ordovices and utterly defeated them in a fierce battle at Anglesey. This expedition had left the remainder of the province denuded of troops, so making it easier for Boadicea's rebellion. The warriors of the Ordovices included bowmen and spearmen.

The Setantii tribe, who lived north of Chester, paid tribute to the Brigantes and were not considered to be a serious threat to Roman rule, while north of Lincoln were the Parisi tribe, who fought mostly from chariots, and might be antagonistic. The Cornovii, who lived in the Midlands, north of Wroxeter, were friendly disposed towards the Romans.

The Coritani and the Catuvellauni were tribes who occupied the forested areas that stretched from north of London up to the midlands, above Leicester. They were secluded people who occasionally raided out from their wooded strongholds. After the rebellion had got under way Boadicea's forces were reinforced by both Coritani and Catuvellaun tribesmen but they represented a belligerent minority rather than the tribes themselves.

Made up of hundreds of undisciplined and independent war bands each under their own petty chief who was similarly intolerant of authority, Boadicea's host was immensely difficult to handle. They advanced in the natural manner of all tribal armies—as a disorderly mob, scouring the countryside for plunder and with little thought for any dangers that might lay ahead. The cavalry and chariots, being fleeter, were in the front while at the rear of the host were thousands of waggons and carts, pack-ponies, donkeys and herds of cattle as it was a tribal habit for Britons to take their possessions with them when they went to war. A large percentage of the force owed her no personal allegiance but were tagging along for loot and a desire to injure the Romans. Boadicea herself travelled in a conspicuously red war chariot drawn by two white horses. It is said that she wore a yellow cloak and golden ornaments to match her hair which flowed unbound over her shoulders. Her two daughters travelled in the chariot with her.

The rebellious Britons descended on the undefended town of Camulodonum (Colchester) where they slaughtered to a man the Roman settlers and their British collaborators. The IX Legion, marching from Lindum (Lincoln) to quell the revolt, were overwhelmed by sheer numbers and almost wiped, out while II Legion's commander, Paenius Posthumus, judged it best to remain in the protection of his encampment at Gloucester. Suetonius Paulinus was engaged, with XIV and XX Legions, in

North Wales and before they could intervene Boadicea's host had burned Londinium and slaughtered its inhabitants. Greatly outnumbered, the two Legions with their auxiliaries stood their ground on a carefully chosen ridge with unturnable flanks, near Staines. The triumphant Britons poured forward to destroy the last Roman force that could oppose them in Britain, advancing in no set divisions without order and formation so that the whole array bubbled and writhed to a background of the bellowing of hundreds of deep-throated war horns, battle cries and war-chants mingled with the neighing of horses and the baying of huge hunting hounds which accompanied them. It is recorded that the Iceni, the largest contingent, formed the centre, with the Coritani on the right and the Catuvallauni on the left. It is not certain where the Trinovantes and Atrebates were positioned. A typical British tribesman wore vari-coloured woollen trousers, bound criss-cross with leather thongs below the knee. His naked body was decorated with blue stripes. The better armed warriors were in the front, forming a broad band of multi-coloured cloaks, brightly enamelled shields and painted helmets. They were tall men, blue-eyed and bearded and with long tawny hair and were almost certainly the best fighters on the British side. A Roman general wrote that he believed their superior stature and physique over that of the Romans gave the western barbarians a false sense of confidence so that, at first sight, they found it hard to believe that they could not overcome the swarthy much smaller soldiers from Rome. In these assumptions, they failed to consider the system and training born of centuries of experience that made the Roman soldier the most disciplined and effective force the world had known up to that time.

The ground in front of the Roman position was soft and sandy so that it easily turned to mire, besides being covered with scrubby vegetation so that the shaggy British ponies and their chariots were reduced to a laboured walking pace or else jammed themselves in a heaving mass on the single causeway that passed through the Roman ranks. The legionaries, six deep, formed a 1,500 yards crescentic line, with cavalry on the wings and with the bodyguard cohort of Suetonius Paulinus in the rear, across the causeway, as the only available reserve. The auxiliary infantry were posted, as skirmishers, in a long straight line some 1,000 yards forward.

Pushed forward by the weight of men behind them, the Britons poured on to the battlefield like flood-water through a broken dyke. After taking casualties from the auxiliary Bracarii slingers and Thracian archers plus short sharp cavalry charges into their flanks, the Britons flung themselves in a headlong charge on the skirmishers who were recalled by trumpet, breaking to the flanks so clearing the Legion front.

A British attack resembled a stampede of wild animals, backed by the pounding of thousands of feet and hoarse war cries and bellows. Flavius Vespasianus, who had led the II Legion in early operations in Britain, considered the tribesmen to fight very well 'for savages' and to be braver than any others he had met—they attacked in mass and did not break easily but were completely lacking in discipline so that they possessed no tactical knowledge that permitted manoeuvring. At Staines the torrent of tribesmen surged forward as the two Legions, with swords held low and shields up, went forward down the slope in a steady controlled run. At twenty paces' range the front rank flung their pila, and then both Legions formed into a series of 'cohort wedges'. With one century at the apex, two more echeloned rearwards on each flank and the sixth at the base of the wedge to support and reinforce where required, it was the ideal battle-formation against such an undisciplined rush of tribesmen. Almost immediately, the wedges were deep in the British ranks, bringing their rush to a standstill as those caught in the 'V's between the wedges were forced to fight on two fronts. The narrow front, incapable of being outflanked, allowed no more than a thousand or so Britons to simultaneously engage the same number of legionaries, regularly relieved by fresh comrades as the battle progressed. The result was inevitable and it was not long before the Britons began to give way until the wedges had flattened and the Roman battle-line was once again a straight line. Choosing exactly the right moment, the auxiliary cavalry charged down the slopes to smash into both British flanks, transforming them into a suddenly dissolving flood that contrasted strangely with the remorseless pushing-back pattern of the long battle-line. Then, with a dramatic heave, the British front broke, turning into a rabble that sought to flee in all directions. But they were trapped in an area bounded by the Romans at one end and, at the other, a barrier formed of their own carts and wagons—in this killing-ground, it is reported that 80,000 Britons died for the loss of about 400 Romans and auxiliaries.

Boadicea poisoned herself and so ended the last British revolt against Roman rule, although in 84 A.D., the Picts, a hardy warlike race who lived in Caledonia (now Scotland), banded together under Calgacus and rebelled against the Romans. With little to oppose the Romans but primitive weapons and courage, the Picts were crushed by the Legions of Gnaeus Julius Agricola at Mons Graupius (probably Mount Killiecrankie in Central Scotland). Although they attacked fiercely, desperately, in the usual headlong charge, the Picts were no match for the legionaries. This battle ended all organised resistance against the Romans in Britain although Hadrian found it necessary to build a 73-mile wall across the country between the Tyne and Solway to keep back barbarian raids.

Finally subdued, the tribes of Britain settled down under Roman rule, many leading families became Romanised and when, in the 5th century A.D., the Roman legions left Britain to defend Rome itself against the barbarian hordes, a black age settled on the country.

## STYLE OF FIGHTING

In true barbarian style, the Ancient Britons fought courageously but without discipline and control in wild rushes made up of dense masses of warriors, lacking order or formation.

Nevertheless, the proficient manner in which the Britons used their chariots—in a wheeling, darting, hit-and-run style—indicates that they had at least a proportion of their forces under some sort of control, with a modicum of discipline and training. On the other hand, it may well be that the nobles who commanded the chariots were literally 'born to them' and were proficient in their use, in much the same way as a tribesman used to hunting with a bow and arrow would use the weapon with equal proficiency under war conditions.

The known leaders of British resistance to the Romans were principally Cassivellaunus (opposed Caesar's landing in 54 B.C.) and Caractacus (who opposed the invasion of Aulus Plautius in 43–50 A.D.). Later Boadicea led a revolt in 61 A.D. which was severely suppressed by Suetonius Paulinus. Although the first two commanders were reasonable leaders of their type, it would be incongruous to classify them as more than 'Average', certainly when opposed to Caesar or Plautius, whom they fought in real life. Boadicea appears to have had little more than nominal control over the army she led in rebellion and their subsequent strategy and tactics would seem to place her in the ranks of a 'Below-Average' commander.

## *Fighting Assessment Chart:* BRITONS

| | *Factor* | *Rating* | *Points* |
|---|---|---|---|
| 1. | Ability to assault in controlled order | None | 0 |
| 2. | Bravery | First Class | 3 |
| 3. | Control | None | 0 |
| 4. | Defensive ability | None | 0 |
| 5. | Desperation | Reasonable | 2 |
| 6. | Discipline | None | 0 |
| 7. | Ferocity | Average | 1 |
| 8. | Manoeuvrability | Average | 1 |
| 9. | Mobility | Average | 1 |
| 10. | Movement, speed of | Reasonable | 2 |
| 11. | Order | None | 0 |
| 12. | Tactical Sense | None | 0 |
| 13. | Training and Drilling | None | 0 |
| 14. | United action, capability for | None | 0 |
| 15. | Weapon Handling | First Class | 3 |
| 16. | Demoralisation, likelihood of | Average | 1 |
| 17. | Impetuous Action, likelihood of | Likely | 0 |
| 18. | Panic, likelihood of | Average | 1 |
| 19. | Reluctance to take definite action | Average | 1 |
| 20. | Unscheduled Manoeuvres, likelihood of | Likely | 0 |
| | | *Fighting Assessment* | 16 |

# THE CAVALRY ARMIES

Leaving their native Steppes the barbarians, mounted on their shaggy little ponies, travelled in vast columns, accompanied by their flocks, horses, carts, women and children. Over mountains and across rivers slowly they traversed Asia from east to west to pour across the frontiers of the old Empires that gave way under their onslaught. To these newcomers of the East the last of the Roman legionaries who had conquered much of the then known world had to surrender their position and power.

The barbarising of the Roman army has already been mentioned. The barbarians who entered the Roman service were mainly mounted warriors and, aided by the steady deterioration of the legionary infantry, gradually they imposed their fighting methods on the Roman generals. In about 300 A.D., Diocletian raised the proportion of cavalry in the army from a tenth to a third. A hundred years later infantry formed less than a half of the total, being relegated to such duties as scouting, skirmishing and garrisoning frontier strongholds. It can be argued that this supremacy of the horse soldier, to last unbroken for another thousand years, was to a certain extent due to the deterioration of infantry during the later Roman Empire. To this must be added the increased fighting efficiency of the horseman because of the use of stirrups. Becoming known during the 6th century, the stirrup made it possible for a mounted armoured man to derive the greatest advantage from his horse. Without the stirrup he could not adequately slash with his sword, thrust with his lance or even use a shield, neither could a mounted archer take proper aim without the steadiness and support of the stirrup. Mounted bowmen were first known in Central Asia—that area of the world where the use of the stirrup was first practised. (*Note:* To the mounted bowman, used to riding and using the bow from a very early age, stirrups would make little difference. His leg grip on the horse would be instinctive and his training in horsemanship and archery would be of an equivalent standard to that of the English longbowman in his hey day. Stirrups made cavalry *shock* effective, but there is still doubt as to when stirrups were first in use. 'Stirrups' of wood or leather may well have been in use earlier than was thought but their perishable nature would leave no trace.) It is claimed that both the stirrup and the short horn composite bow were passed on from the Scythians or the Huns of North China to the Sarmatians of Central Asia and thence to the widely ranging Goths. The latter had both the heavy horses and the stirrups that enabled them to destroy the Roman army of the east at Adrianople in 378 A.D.

For social and prestige reasons, the idea of fighting on horseback appealed strongly to the barbarian aristocracies of Europe and Asia. During the long period from 400 to 1400 A.D. all the great armies of Europe and Asia were formed of mounted warriors backed by weak divisions of ill-equipped infantry. This applied to the armed forces of Justinian and of the later Byzantine Empire, and of the invading hordes of Goths, Huns, Mongols, Saracens and Turks.

Originally pushed westwards by the Nomads who were moving in search of new pastureland, these Asiatic horsemen burst upon the civilised lands around the Mediterranean or into China, raiding, burning, looting but sometimes settling down in a semi-civilised fashion. The Scythians were the first to appear and from about 700 B.C. to 100 B.C. they dominated a restless empire in the grasslands north of the Black Sea. All these Asiatic horsemen were much alike, moving restlessly from pastureland to pastureland, the men mounted on their shaggy little horses, the women, children and all their possessions following in lumbering wagons so that they were called wagon-folk. Their constant companion was the horse, their favourite weapon the bow and arrow and they lived in tents. Unarmoured and wearing light clothing they were noted for their hit-and-run tactics in an energetic and disciplined manner and their cruelty was a byword. On the rare occasons when they went on to the defensive, they formed their wagons into defensive fortresses or laagers.

From Adrianople in 378 A.D. to Crécy in 1346, the horse served man well in war, dictating the course of the majority of the major battles of the period so that the soldier without a horse was regarded as a lesser and inferior being.

## STYLE OF FIGHTING

The Parthian victory at Carrhae (53 B.C.) was a foretaste of the eventual dominance of the Middle East by horse archers. Cavalry mobility and fire-power were shown to be superior to the shock action of the hitherto invincible Roman infantry. It was a victory gained by continual harassment by fire-power in the form of hails of arrows fired by large bodies of horse archers who methodically used up all their shafts before replacing them from a huge reserve carried by a camel train. When the slower-moving legionaries, auxiliary archers and cavalry attempted to pin down the elusive horse archers, the latter fell back and enticed the Romans away from their main body before surrounding them and wiping them out at leisure.

Although they had long dominated warfare in Asia, horsemen had never been able consistently to defeat the disciplined infantry of Greece, Macedonia or Rome. During the first 400 years A.D., the increased use of missile weapons forced infantry formations to be less dense and thus made them more vulnerable to the charges of cavalry. Coupled with this was the slow deterioration of Roman discipline which made it more difficult for the Roman legionary to stand up to a cavalry charge—Adrianople is a prime example.

Saddles with stirrups gave the cavalrymen a firm base from which a lance could be efficiently used and made highly effective through the weight of horse and rider backed by the speed at which the horse was moving. Heavy horses appeared suitable for such shock action and, in time, both horse and rider were covered with coats of chain-mail, making them relatively invulnerable to light missiles or hand weapons. These heavy lancers did not display the mounted archers, rather they complemented each other as the heavy cavalrymen forced the infantry to remain in a close formation so that they provided a good target for the horse archers who softened them up for the final and decisive charge of the heavy cavalry.

*Fighting Assessment Chart:* CAVALRY ARMIES

| *Factor* | *Rating* | *Points* |
|---|---|---|
| 1. Ability to assault in controlled order | Reasonable | 2 |
| 2. Bravery | First Class | 3 |
| 3. Control | Average | 1 |
| 4. Defensive ability | Average | 1 |
| 5. Desperation | Not applicable | 0 |
| 6. Discipline | Average | 1 |
| 7. Ferocity | Reasonable | 2 |
| 8. Manoeuvrability | First Class | 3 |
| 9. Mobility | First Class | 3 |
| 10. Movement, speed of | First Class | 3 |
| 11. Order | Average | 1 |
| 12. Tactical Sense | Reasonable | 2 |
| 13. Training and Drilling | Average | 1 |
| 14. United action, capability for | Reasonable | 2 |
| 15. Weapon Handling | First Class | 3 |
| 16. Demoralisation, likelihood of | Unlikely | 3 |
| 17. Impetuous Action, likelihood of | Rare | 2 |
| 18. Panic, likelihood of | Unlikely | 3 |
| 19. Reluctance to take definite action | Rare | 2 |
| 20. Unscheduled Manoeuvres, likelihood of | Average | 1 |
| | *Fighting Assessment* | 39 |

# THE HUNS

In the 4th century A.D., driving their vast flocks before them, the Huns pushed westward from the Steppes of Asia to appear on the frontiers of Europe, on the shores of the Caspian Sea and flooded into the valley of the Volga. Riding nimble-footed stocky little ponies, these squat tough little barbarians were skilled bowmen. Even today the word Hun is a synonym for savagery and brutality and even contemporary barbarian nations consider the Huns to be a '. . . savage race . . . who made their foes flee in horror because their swarthy aspect was so fearful . . . they are short in stature, quick in bodily movement, alert horsemen, broad shouldered, ready in the use of bow and arrow, and . . . they had the cruelty of wild beasts'. Said to be 'almost glued to their horses' they ate, drank and slept in the saddle.

In battle, they fought with order and discipline, depending upon mobility, surprise, cunning and terror to achieve their aims. They rarely stood up to a determined attack, often feigning a rout in order to encourage the enemy to break formation and boldly chase them, when they would turn and destroy him in his disorder. When forced upon the defensive, they fell back upon their wagon laagers. The Huns were usually well led by experienced leaders who directed the battle, acting upon reports received from subordinates moving with the separate detachments of horsemen to whom their direction of attack and objective was indicated by whistling arrows shot into the air.

Dressed in padded, quilted clothing and hats adorned with fur, the many iron fetishes hanging around their necks and dangling on their chests looked from a distance like a breast-plate made of chain armour. Masters of the in-and-out tactics characteristic of mounted bowmen, the Huns also wielded with equal skill the lasso and lance; in close combat they used swords and often lassoed their opponents.

The power of the Huns reached its peak under Attila whose Asiatic hordes swept into France halfway through the 5th century A.D., murdering, burning and pillaging as they advanced. At the Catalaunic Fields near Chalons in 451 A.D., the Huns lost a decisive battle against the combined forces of Visigoths, Alans and Romans under Theodoric and Aetius (often called 'the last of the Romans'). This the final great victory won by Imperial Rome in the West was a pitched cavalry battle fought on open plains with the infantry acting more or less as spectators. After the initial exchange of missiles, Attila's cavalry charged home—the Romans on the left wing under Aetius held firm but the Alans in the centre gave way allowing the Huns to break through. Wheeling round they took the Visigoths in the rear and in the ensuing mêlée King Theodoric was struck by a missile and trampled to death. Victory was denied the Huns by a charge of the heavy Visigoth cavalry under Thorismond, Theodoric's son, coming down from an eminence on the left flank of the Huns where they had been posted at the start of the battle. Outnumbered, Attila withdrew to his camp between the Seine and the Marne Rivers and, behind the wagon barricade, prepared for the final battle on the following day. But next morning the Visigoths and Roman armies separated and marched away and allowed Attila to escape north across the Rhine to his encampments on the Danube. It has been said that Aetius weighed up the outcome of a Visigoth victory and decided that the balance of power would be better kept with the Hun army still in existence, so he persuaded Thorismond to withdraw and return home to secure his throne.

Although this let-off allowed Attila to make a further raid on

Italy during the following year, it was an important victory because the Huns, never a numerically numerous army, had suffered heavy casualties and a dent in the myth of their invincibility probably deterred other barbarians from joining them as allies. In the following year, Attila died and his loosely organised Empire fell to pieces so that the Huns, as a people, soon disappeared from Europe, probably scattering in the area north of the Caspian Sea.

### STYLE OF FIGHTING

After plaguing and harassing an enemy formation by long-range archery, the Hun light cavalrymen would occasionally close in and finish off the enemy with sword and lasso. But if the enemy could absorb punishment and return it the Hun light horsemen avoided shock action against an organised foe who was capable of withdrawal in good order. Incapable under those conditions of reaching a decisive conclusion, the Hun had to be satisfied with raiding and plundering.

The name Hun has come down through the ages as a symbol of barbarism and terror. The inclusion of a 'Ferocious' or 'Terror' factors should be considered when wargaming.

*Fighting Assessment Chart:* HUNS

| *Factor* | *Rating* | *Points* |
|---|---|---|
| 1. Ability to assault in controlled order | Reasonable | 2 |
| 2. Bravery | First Class | 3 |
| 3. Control | Reasonable | 2 |
| 4. Defensive ability | Average | 1 |
| 5. Desperation | Not applicable | – |
| 6. Discipline | First Class | 3 |
| 7. Ferocity | First Class | 3 |
| 8. Manoeuvrability | First Class | 3 |
| 9. Mobility | First Class | 3 |
| 10. Movement, speed of | First Class | 3 |
| 11. Order | First Class | 3 |
| 12. Tactical sense | Reasonable | 2 |
| 13. Training and Drilling | Average | 1 |
| 14. United action, capability for | First Class | 3 |
| 15. Weapon Handling | First Class | 3 |
| 16. Demoralisation, likelihood of | Unlikely | 3 |
| 17. Impetuous Action, likelihood of | Rare | 2 |
| 18. Panic, likelihood of | Unlikely | 3 |
| 19. Reluctance to take definite action | Rare | 2 |
| 20. Unscheduled Manoeuvres, likelihood of | Rare | 2 |
| | *Fighting Assessment* | 47 |

# THE MONGOLS

Of all the barbarians who spilled out of Central Asia to terrorise and destroy the more civilised nations, the Mongols were the most devastating. So much so that the ravages of the Huns formed but a curtain raiser to those of the Mongols. Greatest of the wagon folk, their disciplined armies overran a great part of the world in the 13th century, extending from Russia and Syria to the coast of China. The Mongols were nomadic Steppe-guerillas of the Northern Gobi, a primitive people inured to hardship and skilled in the weapons and tactics of their characteristic cavalry conflicts. In the early 13th century leadership of the tribe fell to he who was later to become known to the world as Genghis Khan, who became such a great leader, strategist and tactician that he united many warring tribes, jealousies and blood feuds forgotten, into the great Mongol armies that terrorised the world. This military genius who knew the value of discipline, organisation and technology mobilised the prime nomadic qualities of mobility through superb horsemanship coupled with a hardiness that ensured survival in the most primitive conditions plus an undeviating ruthlessness to form an irresistible army.

It was a mobile and disciplined army with a high rate of fire-power. Handled with the superb skill of Genghis Khan it was for many years the best trained, the best disciplined and the best equipped force in the world. A mounted army, divided into light and heavy troopers, following commands given by signal flags, lanterns at night or whistling arrows, it was formed into regiments called Banners. Each Banner consisted of 1,000 men, of 10 squadrons of 100 and each squadron contained 10 troops of 10 men each. Ten Banners made up a touman (a division) and usually 3 toumans made up an army. The Khan's Guard, the best troops, formed a separate touman. Each of the closely knit groups of 10 men lived and fought together under severe disciplinary conditions. Should one or two of the 10 flee from the fight, then the remainder of the group were put to death unless they had fought exceptionally well, in which case only the cowards were killed.

These practices were in keeping with the Mongol policy of employing terror as a deliberate tactic. By no means savages although they were barbarians, their contempt for civilisation was shown by the deliberate and cold-blooded manner with which they destroyed cities, towns and villages. Occasionally cities that had submitted without a fight were spared but those that resisted were razed to the ground and every inhabitant was slaughtered or taken into slavery as an example to other communities.

In ancient times it was generally considered that the Mongol armies were enormous in size because they usually seemed to be in many places at once and invariably equal or superior in numbers on the battlefield. It is now thought more likely that they were fewer in numbers than reputed to be but their commanders were able to affect concentrations and combinations of men because of the Mongol disregard for distance and their speed of movement, contrasting most favourably with the cumbersome, slow in mobilisation and movement, feudal armies of the time.

The Mongol horseman carried a bow in a case on the left hip with a large quiver on the right hip. In addition each man had a long curved sabre carried in a leather scabbard on his back with the hilt behind the left shoulder. Lances, adorned with tufts of horsehair and iron hooks to pull foemen out of the saddle, were carried by some of the riders. But it was the powerful and deadly bow that was their most effective weapon, capable of discharging its shafts far more frequently than the crossbowman whom they often faced. Heavily lacquered helmets of iron or hide strengthened with iron were worn with a leather neck guard studded with iron hanging from the helmet to the shoulders. Breast-plates and shoulder pieces of boiled leather sometimes with iron plates attached were used and small round shields of hide strengthened with iron were carried by some of the lancers. Each horseman carried a felt cape and a sheepskin-lined jacket and possessed at least one re-mount and sometimes as many as half a dozen. Like their masters, the hardy ponies they rode could exist on the barest fodder in the hardest weather. Some of the horses were barded by sheets of hide. In addition to being about the only living thing for which the Mongol had any feeling, the horse also provided its master with sustenance in times of emergency when its rider would open a vein and drink its blood.

The Mongol army formed up for battle in five ranks, the front two ranks consisting of armoured lancers on barded horses and the three rear ranks of archers. There was a considerable distance between each rank and between each formation and the archers moved through the intevals in the front ranks to deliver their fire, taking their places in the rear again before a headlong attack was made. This arrangement of lancers and archers provided an almost perfect combination of fire and shock tactics. Like all barbarian cavalry armies, the Mongols relied on murderous barrages of arrows from their mounted bowmen to harass, confuse and break up the enemy's ranks. At the same time other cavalry formations were encircling the enemy to take him in the flank or rear but the Mongols never closed until the enemy had been weakened and disorganised by their fire. This sweep around the enemy's flanks was a principle of Mongol tactics and invariably succeeded against any but the best disciplined and steady opposition. Although polished to perfection by use and experience, it was a system that called for high standards of discipline, timing and direction.

On occasions, the Mongol fire-power was augmented by portable artillery such as in 1258 A.D. when Hulagu Khan had in his artillery train a large corps of Chinese engineers to work the mangonels and the machines that discharged three large-naphtha-tipped fire arrows simultaneously. The Mongols learned much of military values from the Chinese who, although not warlike, displayed a passion for the art and science of war and at a very early period had developed elaborate tactics and the use of war machines and siege-craft to at high standard.

Terror figured so prominently in the Mongol scheme of things that subjugated Eastern peoples accepted their fate with resignation although they were aware that the fate of captives was to be sent back as slaves to the Mongol homeland by long hard marches in which thousands died or to be flung forward in groups as living shields behind which the Mongol warriors advanced to the assault. Those who survived this tactic were usually killed later when their usefulness came to an end.

Another great Mongol leader was Subutai, a veteran leader who survived Genghis. In 1236 A.D. he invaded Europe with an army said to be as strong as 150,000. Although different from the original Mongol Banners which had been decimated by decades of constant fighting, it was a force of seasoned troops made up from the various tribes of the Steppes plus a number of Chinese—all under Mongol officers and discipline. After slashing through Russia like a plague of locusts, the Mongols rode into Poland and moved towards the plains of Hungary. Decisively defeating an army of Germans and Moravians together with the Teutonic Knights at Liegnitz and causing a Bohemian army under King Wenceslas to turn back, a number of Mongol columns moved in their characteristically rapid manner to join together in Hungary. Opposing them at Mohi in 1241 was a mixed army under King Bela numbering perhaps 100,000 men. With little in the way of missile power to counter the Mongol archery, the battle went in the usual way with furious charges by the Christians being blunted by the wearing and delaying tactics of the Mongols. Before long the Christian cavalry was sadly diminished in numbers and the foot soldiers were becoming disordered by the shower of missiles to which they could not respond. Forced into a retreat which turned into a rout it is said that 70,000 of the Christian army were killed. For some reason the Mongols did not follow up their victories and advance any further into Europe and in 1242 the slant-eyed hordes had vanished back to the East, never to be seen again in Europe. Nevertheless, their coming altered the map of Europe and even today Russia reveals traces of its long domination by the Mongols.

The last had not been seen of the Mongols, however, for the grandson of Genghis, Kublai Khan, who began his reign in 1260, completed the conquest of China. Later a Mongol army attempted to invade Egypt but, in the first major set-back to Mongol arms for nearly half a century, they were astonishingly beaten by a Mameluke leader and his forces from Cairo. With Kublai Khan ruling from the comforts and splendour of the legendary Xanadu, the Mongols began to lose their cohesion and revert back to their previous tribal existence.

In the whole history of warfare, the Mongol soldier must rank high because he possessed so many of the personal characteristics of the perfect soldier—ruthless courage, physical toughness, complete disregard for death, danger and physical hardship, the ability to unquestioningly accept harsh discipline, a professional mastery of his weapons and an unparalleled nobility. Accompanying those factors was a high degree of organisation and first rate leaders whose strategy and tactics were ideally suited for the types of armies they commanded plus the acknowledged fact that the feudal armies opposing them were unwieldy and slow-moving besides being far from united. But perhaps as much as anything it was a question of morale—the Mongol soldiers could not conceive the possibilities of defeat after so many years of successive victories, whilst their opponents were overawed and all too conscious of the penalties of defeat to enter a battle with very much confidence of victory.

## STYLE OF FIGHTING

The Mongol army was one of the best organised, best trained and most highly disciplined armies ever created, led by Genghis Khan, a military genius probably ranking with Alexander, and other generals outstanding for strategical and tactical skill such as Subutai, Chepe Mangu and Kublai.

Disciplined to a far greater degree than any other medieval troops, the individual Mongol soldier was the best trained soldier of his day besides being a hardened and experienced warrior capable of showing the greatest fortitude in the face of all privations.

The style of fighting of the Mongols was based completely on attack, by seizing and maintaining the initiative because of their incomparable mobility which gave them superiority of force at the decisive point. The Mongol horde, moving very fast, advanced in several roughly parallel columns spread across a wide front and behind a screen of light horsemen. When one column encountered the enemy it would hold or retire as the situation demanded whilst the other columns continued to advance and occupy the country to the enemy's flank and rear, usually causing him to fall back to cover his lines of communication. During the enemy's retreat the Mongols would close in, always taking advantage of any confusion in his ranks until they eventually managed to encircle and destroy him. On other occasions, the entire horde would move rapidly away from the enemy to turn and strike him in the flank or rear and sometimes a retreat would be feigned and, having stretched their pursuers and destroyed their cohesion, the Mongols would return to the attack, mounted on fresh horses.

To a high degree, the Mongols had achieved the art of combining missile power with shock effect combined with highly skilled tactical manoeuvres and diversions. Because of their mobility they were able to pin the enemy by a frontal attack whilst another part of their force would swing round to attack him in flank or rear.

The Mongols attacked in five single rank lines with large intervals between each—the first two lines were heavy cavalry and the other three were light horsemen. At the same time other light cavalry units moved in front to screen the attackers and to carry out reconnaissance. Nearing the enemy, the three ranks of light cavalry advanced through intervals in the two lines of heavy horsemen, to dart in and out as they poured upon the enemy ranks a devastating fire of javelins and arrows. So intense was this hail of missiles (the Mongol bow, only a little shorter than the English longbow, was nearly as powerful) that sometimes the enemy were torn asunder without the need for shock action. Otherwise, as soon as they were visibly shaken and confused, the signal was given for the light horsemen to retire and the heavy cavalry to charge home.

Perhaps more than in the case of any other armies mentioned in this book, the Mongols aroused terror in the breasts of all who faced them. This they encouraged by what would now be called psychological warfare as they spread terrifying tales of their cruelty and ruthlessness to those who opposed them. Of course, it was more than propaganda—they really were barbaric and inhuman and frequently herded dense masses of helpless captives in front of them as a shield when advancing towards the enemy or the walls of a city they were besieging. It is essential that a 'Ferocity' factor and its associated 'Terror' factor are used when simulating Mongol warfare. Similarly, it might be considered appropriate to invoke a 'Desperation' factor for opponents on occasions.

Mongol warfare has facets which might well appeal to wargamers who like the unusual. They were masters of stratagems and ruses and disdained no form of trickery which might give them an advantage over their enemy, so that 'Treachery' factors might well be used. They did not go into winter quarters like other medieval armies, frequently using iced-up rivers as roads to reach otherwise inaccessible territories. They had a habit of shooting fire-arrows into camps or towns they were besieging, so causing the confusion they required to cover their advance. Another tactic was the frequent use of fire and smoke screens, caused by detachments setting fire to undergrowth or even using artificial means to provide screens for their movements.

For fuller details of the Mongol military system see Dupuy, pages 340–5.

*Fighting Assessment Chart:* MONGOLS

| | *Factor* | *Rating* | *Points* |
|---|---|---|---|
| 1. | Ability to assault in controlled order | First Class | 3 |
| 2. | Bravery | First Class | 3 |
| 3. | Control | First Class | 3 |
| 4. | Defensive ability | None | 0 |
| 5. | Desperation | Not applicable | – |
| 6. | Discipline | First Class | 3 |
| 7. | Ferocity | First Class | 3 |
| 8. | Manoeuvrability | First Class | 3 |
| 9. | Mobility | First Class | 3 |
| 10. | Movement, speed of | First Class | 3 |
| 11. | Order | First Class | 3 |
| 12. | Tactical Sense | First Class | 3 |
| 13. | Training and Drilling | Average | 1 |
| 14. | United action, capability for | First Class | 3 |
| 15. | Weapon Handling | First Class | 3 |
| 16. | Demoralisation, likelihood of | Unlikely | 3 |
| 17. | Impetuous Action, likelihood of | Rare | 2 |
| 18. | Panic, likelihood of | Unlikely | 3 |
| 19. | Reluctance to take definite action | Unlikely | 3 |
| 20. | Unscheduled Manoeuvres, likelihood of | Rare | 2 |
| | | *Fighting Assessment* | 50 |

# THE SAXONS

The first Saxon invaders to reach Britain were well built flaxen-haired warriors from Germany. Like the Vikings who came later, they were fierce and brutal but very courageous, showing the same sense of discipline and loyalty to the captain of their ship or fleet. Each man was equipped with a seven-foot iron-headed spear and a circular wooden shield bound with iron; he fought in a woollen tunic and his leader wore a mail shirt and wielded a long sword. The first expeditions to hit England were solely for plunder; later came the Saxon settlers, seeking a more hospitable land then the marshes of northern Germany. These successive waves of blond invaders spread their settlements over the southern half of England, they flattened what was left of the Roman houses and camps. Their villages were grouped under the rule of a thegn, usually the man who had commanded the fleet that had brought them to that particular area. Once settled on the land, all men swore to serve the thegn, working his land as well as the land they held under him and marching off at his order in time of war. The law said that all fit males between the ages of 16 and 60 were liable to be called for service in the fyrd for a period not exceeding two months.

At the end of the 8th century the Saxons who had settled in Wessex (the southern part of England from Devon to the Home Counties) began to be sorely troubled by seaborne raiders from Scandinavia. Despite their earlier resemblances to these invaders, softened by their peaceful existence as farmers, the Saxons were nowhere near so organised as the invaders nor so well equipped for war. For more than a century, wave after wave of Norwegians and Danes harried this southern part of England, often coming in fleets of as many as 300 longships. Some Saxon kings organised fleets capable of meeting the Vikings on their own terms. Alfred of Wessex was the most successful of them, organising an army to fight on land. In so doing, he was forced to maintain his thegns and their freemen on a prominent war footing so that they became semi-professional soldiers. In 878 A.D. at the battle of Ethandune Alfred led the forces of Wessex to decisive victory over the Danes under Guthrum so halting and turning back their advance and clearing the Danes from the whole of England south of the Thames and west of Watling Street. In later years, in 991, at the Battle of Maldon in Essex, the Danes broke the Saxon shield-wall and massacred the men of Essex under their Saxon Earl Byrhtnoth.

After many years of bloodshed and fierce fighting, the Danes of northern England became integrated with the Saxons of the south and the country was ruled by a Danish king, Canute, who reigned from 1016 to 1036. Canute went a step forther towards establishing a professional army when he formed a royal guard of huscarls, picked men of the best physique and the greatest skill with weapons equipped with the best arms and mail available at that period. At first the huscarls were all Scandinavians but Saxons were later admitted to their ranks.

As a fighting nation, the Saxons reached their high watermark at Hastings on the 14th October, 1066, where a competent Norman combination of archers and cavalry beat a brave Saxon infantry army in a strong defensive position. On the death of his brother-in-law, Edward the Confessor, Harold Godwinson had been proclaimed King of England. He accepted the throne with the knowledge that William, Duke of Normandy, claimed that it had been promised to him by Edward and was likely to attempt to seize the crown by force of arms. Throughout that summer the National Militia, the fyrd, stood-to, manning a fleet that patrolled the Channel to ward off the anticipated Norman invasion. Like all medieval commanders Harold was beset by the ever present problem of how to keep a militia or feudal army in the field for any length of time so, towards the end of summer Harold reluctantly allowed the fyrd to return to their homes in order to collect in the harvest. It was also necessary to bring the British ships round to the Thames to be refitted and re-supplied.

Then came the news that a Norwegian army led by Harold's outlawed brother Tostig and a famed Scandinavian warrior Harald Hardrada had landed in the Humber. On September 20th the Yorkshire fyrd led by the northern Earls Edwin and Morcar were heavily defeated outside York. On the 24th September King Harold arrived with his huscarls, having covered 200 miles in nine days along the old Roman road—indicating that the force travelled on horses although they fought on foot. At Stamford Bridge on the following day, Harold defeated the invaders, killing both Tostig and Hardrada and allowing only about a tenth of the invading force to get back to their ships.

Hardly had the dust of battle settled on Harold and what was left of his army than the dreaded news arrived that William of Normandy had taken advantage of favourable winds to cross the channel. His force had established a fortified camp near Hastings and were systematically ravaging the surrounding countryside. Harold led his huscarls back to London and after a brief halt they marched out on the 11th October to arrive in the vicinity of the modern village of Battle in Sussex during the night of the 13th October and the early morning of the 14th, having marched 60 miles in 48 hours.

At daylight the Saxon army were in position, drawn up in battle array on a ridge that stretched on both sides of the old road from Hastings to London. They numbered probably between 7,000–8,000 men of all types, less than 2,000 were the huscarls, probably the finest professional soldiers in the world at that time. Having marched from London to York, fought a hard battle before retracing their steps to London and then tramping down to Hastings in less than a month, it is possible that they were physically weary. Dismounted, and in a solid ten-deep formation extending for about 600 yards along the ridge, the Saxons presented an unbroken mass of men and steel; those who had shields had planted them in front to form a shield-wall. Their flanks were secured by a wooded rise on the left and a marshy stream on their right. In the centre of the line, amid the huscarls, waved the Dragon standard of Wessex and Harold's personal banner of a Fighting Man.

The Saxon army was an infantry force with a nucleus of trained professional soldiers who were equipped in very much the same manner as the Normans and wearing interlocking link chain mail shirts (byrnies) over heavy leather or quilted cloth garments. Their helmets had metal strips (nasals) running down over the bridge of the nose and their large kite-shaped shields were made of wood reinforced with iron. Backing them were the theyns, men of wealth and standing who had some military training and were reasonably well armed and equipped. But the majority of the force consisted of the fyrd or militia of southern England, peasants with swords, spears, javelins, clubs and primitive weapons made from agricultural implements, and lacking any protective clothing. The huscarls wielded a two-handed axe that could bring down both man and horse with a single blow; the theyns used long straight swords, spears and light lances. The Saxons were brave and tough fighting men but, with the exception of the huscarls and some of the theyns, they were completely undisciplined—which was to have a decisive effect on the outcome of the battle.

Harold realised that William's archers and armoured infantry presented a problem but hoped to counter the Norman cavalry by his continuous line of shields, spear points and axes along the crest of the ridge. William opened the battle by sending his archers forward up the hill until they came within range of the Saxon line and both sides exchanged fire. Lacking missile-men, the Saxons hurled axes and javelins. Because they were firing uphill against men protected by shields, helmets and mail shirts, the Norman fire was not very effective and soon the Norman infantry passed through the retiring archers and mounted the slopes to assail the whole Saxon line. The superior weapons, physical skill and sheer courage of the Saxons heavily defeated the Normans all along the line. William sent in his heavy cavalry but they were disordered by their own infantry and slowed by the steepness of the hill so that they slowly approached the Saxon line throwing their lances like javelins before engaging individual duels with the huscarls. On foot, in close formation and armed with long axes, the strong foot soldier had an advantage over the single mounted man so the cavalry attack was beaten off.

In the centre and on the right the Norman cavalry fell back in good order but on the left wing the Bretons panicked and fled through the swampy ground in confusion. In a wild and undisciplined pursuit, Harold's entire right wing chased down the hill after them. Rallying his men, William wheeled his centre and led the Norman cavalry against the pursuing Saxons. Caught in the open, they were cut to pieces and Harold had lost one fifth of his entire force in two or three minutes. Now followed a series of desperate uphill cavalry attacks on the gradually shortening Saxon front, each one beaten back. It is said that William deliberately repeated the manoeuvre of the left wing by ordering his right wing to feign panic and draw the Saxons down the hill after them, to be cut down as before.

The day wore on and the Saxon army, despite losing almost the whole of both wings, held firm with the huscarls closing their ranks around the standards. Desperately seeking a way to break their line, William began to alternate cavalry charges with showers of arrows so that when the horsemen were moving forward the bowmen covered their attack by loosing arrows high in the air to fall from the sky into the Saxon formation. Towards the end of the autumn day, and with both Harold's brothers killed, the Saxons became too few in number to occupy the whole of the ridge. This permitted the Norman cavalry to use the smooth clear top of the hill for shock charges whilst their archers fired at ranges of less than 25 yards. These cavalry charges from the flanks pushed Harold's surviving huscarls too close together to efficiently wield their four foot axes. Harold himself was badly wounded when hit in the face with an arrow but was able to continue in command of his rapidly diminishing army. As dusk approached, the Saxons occupied only the middle third of their original position with both flanks exposed.

Harold was almost blind but with shield and axe fought on in the midst of his men. On his fourth horse of the day, the first three having their skulls cleft by axes, William led an all-out charge into the area of the banners. Harold was killed and the English formation finally broke, fleeing on all sides to be slaughtered by Norman cavalry as they ran.

Hastings showed that under certain conditions infantry could hold a defensive position against cavalry but it also revealed that the weakness of infantry during this cavalry period was their inability to take the offensive. Only highly disciplined and well drilled troops could do this and the disorderly and impetuous Saxon downhill charges were their downfall. At the same time, unaided by archers, the Norman knights were powerless to break the infantry line. A victory of feudal cavalry over old-fashioned formed infantry, Hastings pointed the lesson that even the best of infantry in a favourable defensive position, lacking cavalry support, could be beaten by a general of William's ability who knew how to combine horsemen with archer. It was to be the pattern of field warfare for the next 300 years.

### STYLE OF FIGHTING

The Saxons employed much the same kind of disorganised infantry tactics as the Franks, assaulting wildly and bravely in crowded orderless masses which hit the enemy line with great shock effect.

The well trained, disciplined and brave huscarls formed the nucleus of the later Saxon armies whilst the fyrd was an undisciplined, ill-trained and badly controlled militia who fought, man for man, as an orderless mob.

During the 7th–9th centuries in Britain it is possible that the art of war sank to a lower level than anywhere else in the western world. Strategy, tactics and discipline were unknown disorderly mobs of opposing warriors formed in roughly parallel formations moved towards each other until they met in a hostling, sprawling mêlée that went on until one side broke and fled, to be butchered by their pursuers.

*Fighting Assessment Chart:* SAXONS

| *Factor* | *Rating* | *Points* |
|---|---|---|
| 1. Ability to assault in controlled order | Average | 1 |
| 2. Bravery | First Class | 3 |
| 3. Control | Average | 1 |
| 4. Defensive ability | Reasonable | 2 |
| 5. Desperation | Average | 1 |
| 6. Discipline | Average | 1 |
| 7. Ferocity | Not applicable | – |
| 8. Manoeuvrability | Average | 1 |
| 9. Mobility | Reasonable | 2 |
| 10. Movement, speed of | Reasonable | 2 |
| 11. Order | Average | 1 |
| 12. Tactical Sense | Average | 1 |
| 13. Training and Drilling | Average | 1 |
| 14. United action, capability for | Average | 1 |
| 15. Weapon Handling | Reasonable | 2 |
| 16. Demoralisation, likelihood of | Average | 1 |
| 17. Impetuous Action, likelihood of | Likely | 0 |
| 18. Panic, likelihood of | Average | 1 |
| 19. Reluctance to take definite action | Average | 1 |
| 20. Unscheduled Manoeuvres, likelihood of | Likely | 0 |
| | *Fighting Assessment* | 23 |

# THE VIKINGS

Erupting from the cold northlands of Scandinavia in the 9th and 10th centuries, these warriors were among the most ferocious and warlike races of all times, arousing fear and confusion throughout Europe, even as far south as Gibraltar and east as the Caucasus. From 787 when they attacked part of the Dorset coast, when Harold of England defeated them at Stamford Bridge in 1066, these Norwegian and Danish pirates harried the southern half of England, their black-tarred, dragon-headed longships penetrating far up the waterways that led inland from the sea. Nowhere was safe, the black ships probed up rivers and creeks, negotiated the most treacherous coasts and even ploughed into the middle of guarded harbours. Terrified villagers fled to the nearest town for safety and beacons flamed on the headlands at the sight of the striped sails and low hulls of their longships.

By the middle of the 9th century the number and scale of their raids on the lands to the south had so increased that they were regarded as a seasonal affliction. Like the devouring, irresistible flames of a forest fire, they darted out from Scandinavia to terrorise Finland, Lapland, Iceland, the Orkneys, the Shetlands and the Hebrides. They left their mark on Ireland, Scotland and England; Germany and Russia knew them; they advanced by the Garonne as far as Toulouse and penetrated into Spain where they fought battles with the Saracens. Mediterranean France and Italy suffered their attacks as the Vikings sailed between Spain and Africa and entered the Rhône to establish themselves in the Camargue. They went to Italy and sacked Pisa among other cities, whilst a force is known to have fought in Galicia. Many of them, such as Harald Hardrada, were in the service of the Emperors of Byzantine. Their smaller raiding parties avoided walled cities but, when in sufficient strength and if they considered the prize worthy of the effort, they were even known to have undertaken sieges—one such attack on Paris lasting from 885 to 887.

Viking ships were long, narrow, flat-bottomed vessels that rose steeply at bow and stern; being over 25 paces long and about a third of that wide, they looked like an elongated walnut shell. The bow had a small deck and reared up into a carved dragon's head; at the decked stern a tail was carved. Their square sails were painted or made up of lengths of coloured woven cloth, but they could be propelled by oars. One well known longship, the *Long Serpent* was probably 150 feet long with 34 rowing benches giving 68 oars. The spaces between the benches were called rooms and in action the *Long Serpent* had 16 men in each room. With the defenders at stem and the low after-deck this famous ship carried a total of about 600 fighting men—but it was a very much larger vessel than most. Their very narrow keels enabled the longships to scrape their way up a shallow stream or to ride the huge breakers of the open sea and with tree trunk rollers placed under them, their tough crews could push them overland from one river to another.

Holding undisputed control over the seas of Northern Europe for more than two centuries, almost all the great sea battles the Vikings fought were against each other. With the exception of Alfred of Wessex, there seems to have been few organised attempts to raise a fleet with any hope of success against the Vikings. In spite of their great reputation as seamen the Vikings appeared to have made little use of seamanship in battle, the high curved stem was ill-suited for ramming and little attempt was made to maintain any tactical formation although the longships seemed to have been grouped in fleets or squadrons. The Vikings flung their ships at each other in a wild and terrifying onslaught in the same way as they fought on land. It resembled land fighting even more when two or more ships were lashed together in a fighting platform enabling the fighting men to pass from one to the other. Grappling hooks and anchors were used to draw the opposing ships together and hold them fast so that the Vikings could surge on to the enemy's decks in a frightening and irresistible flood.

The abundant seas and waterways provided the Vikings with their avenues of attack, enabling them to land their raiding parties wherever they desired with unrestricted freedom of action. Because a Viking attack could be launched anywhere where there was sufficient depth of water to float a longship they possessed an immense mobility and an almost insuperable advantage over the surprised defenders. By the time local leaders had gathered together sufficient men to make a stand the pirates were away, driving their prisoners before them and leaving behind burning buildings and dead bodies. Then men defending their homes and families against the Vikings were invariably outclassed because they lacked the ferocity of the pirates and fought with inferior arms. The Vikings carefully selected their victims from those whom they knew to be weakest and avoided pitched battles—sound tactical reasons why they were not contained until such leaders as Alfred turned their men into semi-professional soldiers by maintaining them on a permanent war footing so that they capably manned large groups of warships or massed as armies to attack the invaders sometimes overland on horseback.

In time, their raids became more planned, being preceded by signallers and spies who informed them of the districts where resistance was weak and prospects for loot were good. Sometimes they seized an off-shore island to provide a jumping-off point for future operations or a winter base. As the years of raiding went by, the inhabitants of coastal areas abandoned their land and villages to move inland into fortified towns, but the Norsemen, unwilling to be defeated by the vast stretches of deserted coast lacking loot and slaves, scoured the countryside for horses and sent mounted parties of raiders far inland to spread devastation amid communities which had considered themselves to be safe. The Vikings did not always come for loot. Sometimes, as in Normandy, they sought land and through fear-inspired treaties they carved themselves out empires in lands more hospitable than the frozen north from which they originated.

Although the term Vikings is used to cover all the raiders from Scandinavia it should be understood that there was almost constant fighting between the Norwegians, the Danes and the Swedes. Lacking any sort of national feeling and bound only by ties between kinsmen or neighbours, every man's hand was raised against those around him and Dane had no qualms about fighting Dane nor did Norwegian abhor robbing a fellow Norwegian. On those rare occasions when planned campaigns were undertaken, armies were made up of large groups called fylkings, formed of localised detachments (sveiter) led by their own chiefs—the numbers of men in these groups varied considerably. Each fylking had its own standard and fought in a wedge-shaped formation with the best warriors at its apex.

Everything about the Viking, from his reputation to his appearance, terrified his opponents, who saw him more as a wild animal than a human creature. Truly, they were a fearsome sight, dressed in wolf-skins and sheep-skin breeches, short-sleeved shirts (byrnies) made of leather covered with overlapping plates or scales of iron or chain mail; on their heads they wore great iron helmets sometimes adorned with horns or raven's wings. Huge axes, very wide in the blade and long-handled, stuck in their leather belts—in battle they were wielded with both hands. Also, they used swords and spears. Often his most cherished possession, the Viking's sword was either a straight two-edged weapon with a blade nearly 3 foot long or a shorter 24 inch broader bladed single-edged sword. Their spears were of various types, some resembled halberds while others were heavy thrusting weapons, then there were lighter, barbed spears for throwing. The bows used by Viking archers were of elm or yew and about 5 feet in length with a shaft from 2 to 3 feet in length so that the bow was probably drawn to the ear as in the case of the English archers 300 years later. Shields were round and made of wood with an iron rim and boss and were painted, sometimes with an intricate design. At sea the shields made a brave show and protected the rowers by being fixed along the gunwales but they were never carried this way when the ship was under sail because they would have been washed away.

The Viking fighting unit of captain-and-crew operated efficiently through its bonds of comradeship and discipline. Invariably both were stronger than in the enemies they opposed. These organisational qualities added to their exceptional mobility often gave them superiority in numbers at the vital moment of contact while their ferocious fighting qualities, backed by ruthlessness and savagery, gave them a moral superiority that constantly increased as their victories multiplied. But there was more than this that brought success to the Vikings—for example they are said never to have camped without first erecting fortifications and were known to have used traps so that opposing forces would suddenly plunge into deep ditches covered with leaves and branches.

The Vikings possessed an almost legendary ability in the personal handling of their weapons, using both hands with equal faculty—they had a trick of suddenly switching the weapons from one hand to the other to throw an opponent off balance. Some of them possessed enough strength to cast their spear right through an enemy while others were sufficiently keen-eyed to catch an opponent's spear in mid-flight and hurl it back at him. Shields were used not only for protection but as weapons of attack—an opponent was disarmed by a quick twist and then brained by a sideways sweep with the edge of the shield. Another favourite Viking trick was to allow an enemy's weapon to imbed itself in the shield and then jerk its holder to the ground, striking him on the neck with the shield rim as he floundered at the Viking's feet. Two Vikings, each guarding the other's back, were quite happy to take on three or four times their number. They slashed rather than thrust with the sword at an enemy's shoulder, knee or ankle—the more agile opponent jumped over the sword but if he mistimed his leap he would be speedily finished off whilst off balance. The Viking's opponents sometimes tried to counter spear thrusts at close quarters by jumping and straddling the weapon as it came at them or by trying to split the shield by a single axe blow down the middle, exposing its holder to a deadly following stroke. A type of axe was evolved with a blade that beat down the shield, and a spike that was used with a lever movement to pierce the shield bearer's chest or throat.

The Saxons knew that the only way they could stand against a Viking onslaught was to form a shield-wall, using their bucklers to make a barrier from within which they fought. This tactic was countered by the Vikings crumpling the shield-wall with javelins thrown from a distance until it was broken and the battle ground became a mass of separate combats. The shield-wall was the only effective counter to the basic Viking plan of attack when they made a fierce frontal rush accompanied at the same time by a swiftly wheeling flank attack so driving one half of their opponents on to the other half and causing chaos in their ranks. It was always a source of comfort to have horses tethered nearby in case the need arose to escape!

Although the Viking knew better than most the need for order and cohesive action in battle, he was very attracted by 'the individual champion' so that these 'supermen' played a role in their communities rather resembling the 'fastest draw' gunmen of the old American West. Even more dreaded than the champions were the berserks, warriors who were overcome by a sort of fit at the sight of the enemy so that they ripped off their protective armour and then rushed wildly forward in a solo attack. Berserkers were regarded with awe and terror because it was well known that once a berserk had begun his charge there was no going back, by stripping off his armour, the berserk showed a complete disregard for his own safety, indicating that he intended to bring down as many of the enemy as possible before he was himself killed. Obviously, berserks were considered to be formidable opponents and most leaders liked to have one or two warriors from well known berserk families in their ranks. With two groups of shouting warriors facing each other, waving their weapons to intimidate the enemy, suddenly the berserk would leap away from his comrades and rip off his mail shirt, then followed a performance of howling, biting the rim of his shield, dancing and capering in a spectacular manner that aroused awe and dread in both friend and foe even before he began the wild charge that sometimes carved its way deep into the enemy ranks before he was chopped down.

Living only for war and spurred on by greed and a complete disregard for both their own lives and those of their enemies, the Vikings represented a dread flood-tide of horror that ebbed and flowed without ever being more than momentarily checked for more than three centuries. History probably heaved a sigh of relief when they succumbed to the attractions of a life more peaceful than any they had previously known and settled permanently in the fertile province of Normandy.

## STYLE OF FIGHTING

The Vikings were skilful warriors who loved fighting and usually had a high standard of discipline when acting in small groups because it was based on loyalty to their own immediate chieftain.

Very mobile and manoeuvrable (particularly in small groups) they had an edge over their opponents because of the terror they aroused in them. Thus 'Ferocity' must be an essential factor when simulating their style of fighting, and the 'Desperate' factor must be considered for their opponents. In action, the Vikings first confused and demoralised their opponents by a hail of missiles—javelins and arrows. When they considered that the 'shield-wall' (the usual and perhaps only effective form of defence against them) had been shaken, then they came in a fierce frontal rush that was accompanied by a swift wheeling movement to a flank so driving one half of a force on to the other and breaking their cohesion. Once this had been done, the ferocity and fighting ability of the Vikings was usually quickly effective against their more numerous but poorly armed, ill-trained and badly led peasant or militia opponents.

Over and above the 'Ferocious' factor inherent within the Vikings themselves must be considered the even more terrible berserks. It is suggested that, within the framework of wargaming rules, allowances be made for these men and their frightening powers. The Vikings also had a liking for champions—the individual warriors who came forward from the ranks to challenge anyone from the opponent's force who fancied his chances.

*Fighting Assessment Chart:* VIKINGS

| | *Factor* | *Rating* | *Points* |
|---|---|---|---|
| 1. | Ability to assault in controlled order | Reasonable | 2 |
| 2. | Bravery | First Class | 3 |
| 3. | Control | Reasonable | 2 |
| 4. | Defensive ability | Reasonable | 2 |
| 5. | Desperation | None | 0 |
| 6. | Discipline | Reasonable | 2 |
| 7. | Ferocity | First Class | 3 |
| 8. | Manoeuvrability | First Class | 3 |
| 9. | Mobility | First Class | 3 |
| 10. | Movement, speed of | First Class | 3 |
| 11. | Order | Reasonable | 2 |
| 12. | Tactical Sense | Reasonable | 2 |
| 13. | Training and Drilling | None | 0 |
| 14. | United action, capability for | Reasonable | 2 |
| 15. | Weapon Handling | First Class | 3 |
| 16. | Demoralisation, likelihood of | Unlikely | 3 |
| 17. | Impetuous Action, likelihood of | Rare | 2 |
| 18. | Panic, likelihood of | Unlikely | 3 |
| 19. | Reluctance to take definite action | Unlikely | 3 |
| 20. | Unscheduled Manoeuvres, likelihood of | Rare | 2 |
| | | *Fighting Assessment* | 45 |

# THE NORMANS

In the beginning of the 10th century a Norseman named Rolf led his men into France where, after behaving in the usual violent manner of Norsemen in other people's countries, he was induced to settle by being baptised and marrying the daughter of the Frankish King, Charles the Simple. In 911, a treaty was signed that turned Norsemen into Frenchmen. They called their new country Northmandy, the land of the Northmen (later the name was corrupted to Normandy), and here the exiled Vikings built up a military state based on the law that men could be called up for service under conditions based on their holding of land.

As time passed, this mixture of Franks and Northmen produced a breed noted for toughness, courage and skill with arms that far surpassed the native Franks in their ability to fight from horses. These Normans had much of the wildness of their northern forefathers, tempered by the cunning and restraint of the natives with whom they had interbred. They travelled abroad and conquered in Sicily, Southern Italy, Asia Minor and the Holy Land and, greatest of all, in Saxon England. Naturally turbulent, they displayed all the old Viking instincts of robbery and violence but now it was from a stone castle or keep instead of a longship. Their inclinations towards anarchy were held in check by their acceptance of the feudal system by which land was leased to barons on condition that they served the king in war. Each baron divided his territory among his knights and in return the knights swore to take up arms when called upon by their lord. In their turn, the people living on the knight's manor held their lands on the same terms of service.

In the middle of the 11th century, William the Bastard had become Duke of Normandy after a hard struggle to gain full control of his duchy. Once in the seat of power, William ruled with an iron fist, showing in numerous battles with his neighbours of Anjou and various rebellious barons that he was a brave and accomplished soldier. In 1066, he began his preparations to assert his claim to the throne of Saxon England, gathering together an army that was not an array of feudal tenants but a feudal force in the sense that its nucleus consisted of mounted and mailed knights. The tenants 40 days' service per year military obligation was no use at all for an overseas expedition so William raised his army by voluntary recruitment, attracting military adventurers from all over Europe to his banner with alluring promises of land and booty. Attracted for the same reasons, his Norman barons followed him voluntarily bringing with them forces of spearmen and archers.

The weapons and armour used by the Normans at the time of Hastings were much the same as that worn by the rest of Europe and consisted of a thigh-length shirt that reached down to the calves, split at the back and front like a divided skirt to make riding possible, covered with rings or overlapping plates to form chain mail. The top part was made as a coif or hood to cover the head and shoulders, surmounted by a crestless conical helmet with an iron strip (nasal) to guard the nose from a sword slash. The lower limbs were usually wound with strips of cloth or soft leather although the Bayeux Tapestry shows some men wearing what appears to be leg coverings of mail. The wooden shields bound with iron were kite-shaped. The knight might carry a ten-foot-long lance bearing a pennon with from two to five points—these lances were sometimes thrown like javelins. Their sword was long, uncurved and double-edged with a straight cross-guard and a pommel terminating in a round knob. Some Norman knights carried a mace, usually hung at the left side of the saddle-bow. The foot soldiers, archers, spearmen and swordsmen, were also armoured so that they could carry out their role of advancing and thinning out the enemy before the cavalry attack. The Bayeux Tapestry shows the bowmen, who made up the majority of the Norman infantry, to be unarmoured and without headgear.

At Hastings the Normans, under the firm command of a leader like William and reasonably well trained for their day, were capable of a certain degree of concerted action, indicated by the 'feigned' flights of their cavalry that induced the undisciplined Saxons from the shield-wall so that they could be cut down. This degree of command and discipline had the greatest bearing on the result of the battle where the heavy cavalry of the Normans were offset by the strong Saxon position on the ridge and by the quality of Harold's huscarls, small as they were in number. Perhaps it was the archers that tipped the balance in the Norman favour. William knew about the long English battleaxe and the strength and valour of the huscarls; he knew that these men could present a line of shields, spear points and axes which could stop his cavalry so he reasoned that his hopes of winning lay in shaking the Saxons' defensive formation with arrows and then crushing them by the shock charge of heavy cavalry. William probably brought to the battlefield 1,000 archers, 3,000 cavalry, and 4,000 heavy infantry, divided into three divisions.

(1) On the left, the Breton Division commanded by Alan Ferquant of Brittany (about 2,100).

(2) On the right, the Franco-Flemish Division commanded by Roger of Montgomerie (about 1,600).

(3) In the centre, the Norman Division commanded by Duke William with his half-brothers Robert, Count of Mortain and Odo, Bishop of Bayeux. This was the larger division and numbered about 4,300.

Each division was in three lines, archers in front, heavy infantry in the second line and mounted knights and men-at-arms in the third.

A good commander, William was able to successfully combine his cavalry and his bowmen, alternating showers of missiles with sudden cavalry charges—a military tactic still in use as many years later as Waterloo. Even the best of infantry, lacking fire-power and without cavalry support, were helpless against such tactics but neither could the Norman knights, *unaided by archers*, break the Saxon infantry line. In fact, the Saxons would probably have held their own but for the foolish and impetuous downhill charge of first their right- and then left-flank forces. The later Norman conquest of Ireland showed that a mailed and mounted knight could hack down almost any number of half-naked Celtic kernes, but against armoured infantry the issue was not quite so straightforward.

Harold was defeated at Hastings because Saxon courage and doggedness could not stand against Norman fighting ability and courage, backed by cunning. The qualities displayed by both Saxon and Norman showed to great advantage on many later battlefields when the two nations had become welded into one.

## STYLE OF FIGHTING

Like most armies of the period, the Normans were predominantly a cavalry force although, being subject to the feudal system, their serfs and peasants fulfilled their military obligations by acting as ill-trained and poorly armed militia. On the other hand, the nobility were men trained from birth to the saddle and the use of arms, were reasonably capable soldiers although difficult to control and lacking in discipline. It will probably be necessary when wargaming to utilise the 'Feudal' factor.

The general style of fighting of the Normans appeared to consist of a preliminary 'softening-up' by archers and missile men, followed by a headlong charge of heavy cavalry. On occasions knights and men-at-arms would be dismounted, either because the terrain was considered more suitable for this tactic or else as a means of stiffening up the unreliable and poorly trained feudal levies. Also, so dismounted, the knights and men-at-arms, backed by the type of discipline given them by their chivalrous codes of honour, stood a better chance of withstanding a cavalry charge than did the levies.

During the Hastings campaign, William the Conqueror used a considerable number of mercenaries, thus a 'Mercenary' factor should be included within the framework of wargaming rules for this period. William, and to a certain extent Harold, could be considered, in this period, to be 'Above-average' commanders.

*Fighting Assessment Chart:* NORMANS

| Factor | Rating | Points |
|---|---|---|
| 1. Ability to assault in controlled order | Reasonable | 2 |
| 2. Bravery | First Class | 3 |
| 3. Control | Reasonable | 2 |
| 4. Defensive ability | Average | 1 |
| 5. Desperation | Not applicable | – |
| 6. Discipline | Average | 1 |
| 7. Ferocity | Not applicable | – |
| 8. Manoeuvrability | Average | 1 |
| 9. Mobility | Reasonable | 2 |
| 10. Movement, speed of | Reasonable | 2 |
| 11. Order | Average | 1 |
| 12. Tactical Sense | Average | 1 |
| 13. Training and Drilling | Average | 1 |
| 14. United action, capability for | Average | 1 |
| 15. Weapon Handling | First Class | 3 |
| 16. Demoralisation, likelihood of | Rare | 2 |
| 17. Impetuous Action, likelihood of | Average | 1 |
| 18. Panic, likelihood of | Rare | 2 |
| 19. Reluctance to take definite action | Average | 1 |
| 20. Unscheduled Manoeuvres, likelihood of | Average | 1 |
| | *Fighting Assessment* | 28 |

# THE BYZANTINES

Through nearly five centuries of almost constant victorious warfare the relatively small professional army of Byzantium, best in the world because of its organisation and leadership, held the eastern entrance to Europe against the onslaughts of Persians, Arabs, Turks, Bulgars and Slavs. Aided by a fine fleet, these armies of the Eastern Empire pursued the craft of war in a scientific and disciplined manner, using flexible formations with well devised tactics and strategy, backed by logistical and engineering methods far in advance of anything conceived by their enemies. Because they were almost invariably outnumbered, the Byzantine armies had to be formed of the right combination of arms and tactically employed to best combat the anticipated methods of specific opponents. Most of these enemies were horsemen and the operations usually defensive so that the Byzantines, profiting from the experiences of the defenders of the old Roman Western Empire, whose armoured infantry units moved too slowly to bring mounted invaders to action, were themselves predominantly cavalry capable of rapid concentration.

In the early days of the Eastern Empire the armies were similarly composed to those of the West, largely of foreign auxiliaries. In addition to obvious problems of loyalty, this method was expensive and uneconomic. Justinian, Emperor of the East (527–565), determined to re-conquer the West and in a series of campaigns his brilliant generals, Belisarius and Narses, seized Africa from the Vandals and Italy from the Goths. Belisarius relied almost entirely on his cavalry and is said to have learned their use from the Goths themselves who were defeated by the superiority of the Byzantine mounted bowmen over the Gothic foot archers.

By this time most armies, whether barbarian or Roman, depended primarily upon cavalry. As early as the 3rd century A.D. heavy armoured cavalry, copied from those of the Sassanian Persians (successors to the Parthians), had been introduced into the Eastern Roman army. In 538 A.D., Belisarius took an army into action formed of 4,500 horsemen and only 3,000 infantry. Many of these foot soldiers found horses for themselves and took the field as light cavalry in order to get a share of the fighting and loot. It is said that, following a remonstrance from his Isaurian archers, Belisarius positioned them in the forefront of battle where they were eventually ridden down by their own cavalry galloping through them. Belisarius himself wrote:

> Our own regular Roman horse . . . capital horse bowmen, while the enemy has hardly any knowledge of archery. For the Gothic knights use lance and sword alone, while their bowmen on foot are always drawn up to the rear under cover of the heavy squadrons. So their horsemen are no good till the battle comes to close quarters, and can easily be shot down while standing in battle array before the moment of contact arrives. The foot archers, on the other hand, would never dare to advance against cavalry, and so keep too far back.

In addition to mounted bowmen, Belisarius employed heavy cavalry armed with lances, setting them in a solid body to await the enemy charge with archers deployed on both flanks—into this three-sided box the less disciplined Goths came to destruction in a manner that foretold Crécy and Agincourt.

Similarly, Belisarius's successor, the eunuch Narses, foreshadowed the English tactics of the Hundred Years War at Taginae in 552 A.D. Here Narses dismounted his non-Roman lance-bearing cavalry and placed them in phalanx-like order in the centre of his battle formation. On either wing, he angled about 4,000 Roman foot archers so that the whole force was in half moon formation. Behind each wing of archers was placed a strong force of Roman cavalry and well to the left two more strong bodies of cavalry were held in reserve. As the Gothic heavy cavalry charged forward they were slowed down by the archers on the flanks until those who reached the densely packed lancers in the centre were moving too slowly to make any impression upon them. After a number of charges that were all treated in this fashion, the Goths began to retreat only to be hit by the Roman cavalry swinging into their flanks, causing the demoralised Gothic horsemen to retreat back through their own disordered infantry. A year later, Narses had a similar success in Italy against the Franks.

Emperor Maurice (582–602) re-formed the Byzantine army, transforming it into the type of army which, with little change, was to hold off the Eastern invaders for almost 500 years. The heavily armed and armoured cataphracti, armed with sword, dagger, bow and lance, emerged as the elite core of the Byzantine army. Helmeted in steel with a nodding crest and wearing a long shirt of mail from neck to thigh, gauntlets and steel shoes with a surcoat worn over the mail, these heavy troopers rode charges protected by frontal armour. From his lance streamed a long pennant bearing the device of his unit—each had distinctive coloured crest, surcoats and pennants. A light cavalry trooper, armed with a lance and sword, carried a large shield but wore less armour. Playing a subordinate role, the infantry were composed of heavy and light troops—scutati or heavy infantry wore crested steel helmets and a short hauberk and carried a large shield bearing, like their crest, the colour of the regimental banner. They were armed with a lance and sword and a heavy spiked battleaxe. The light infantry were archers with bigger and more powerful bows than those carried by the cataphracti. Every four cavalrymen had an attendant and every sixteen infantrymen were attended by a man who drove a cart carrying their heavier gear. There were ambulance corps with doctors and stretcher bearers and a well organised supply corps.

The Byzantine army was divided into tactical units of 300–400 men called the tagma, 3 or more tagmas made a small brigade and 3 brigades a turma. These formations were tactically used in a manner best suited to defeat the enemy opposing them. If fighting heavy cavalry such as the Franks, the Byzantines held off and lured them into unsuitable country: Against light mounted bowmen, the tactics were an immediate charge, as they could not stand against the cataphracti nor could they withstand the Byzantine foot archers whose heavy bows out-ranged their own.

Byzantine cavalry were usually drawn up in three lines with flanking troops on the wings and two units well out on the flanks of the enemy whose role was to charge into flanks or rear if the opportunity offered itself. Basically, it was a formation designed to allow as many successive cavalry attacks as possible to be made. When infantry were present they were usually stationed in the centre of the line with the cavalry on both wings. The Byzantine archer-cavalryman combination allowed the enemy to be assailed by a devastating combination of fire-power and shock so that formations decimated by arrows were then ridden down by successive charges of heavily armoured horsemen. It is reported that at Silistria in 941 a Russian force of 60,000 infantry, probably many of them being Vikings or Norsemen, were defeated by a mixed force of 30,000 Byzantine infantry and cavalry. The arrow storm of the Byzantine archers decimated the great shield-wall formations of the Russians until their thinned ranks were ridden down by armoured cavalry.

It is doubtful whether Edward I, Edward III or Henry V were ever aware that the tactics they devised to demoralise the French had been highly effective some 400 years before!

The later Byzantines believed the most wily and dangerous enemy they ever had to face were the Saracens, who were not only more mobile than the armies of the Eastern Empire but were immeasurably superior in number. In the 7th century when first they poured out of the Arabian Desert, the fanatical zeal of the Saracens made them almost irresistible and even the well trained and disciplined Byzantine armies were unable to cope with them. For centuries the Byzantine army repulsed sporadic Saracen raids into Asia Minor, often destroying these parties when they were retreating, plunder laden, rather than attempting to intercept their fast-moving parties at the outset of their raid. In time, the sole advantage of the Saracens lay in numbers, speed of attack and fanaticism, but when actual contact was made they could not stand up to the Byzantines who, horse for horse and man for man, were heavier and could ride the Arabs down. However, the Saracens got close enough to besiege Constantinople the capital of the Eastern Empire in the 7th and again in the 8th century but were unable to take the city because of its strength and the ingenuity of the Byzantine engineers plus the redoubtable defence put up by the army and the fleet. Constantinople was the most formidable fortress in the world at that time, protected on the seaward side by an eight-mile-long causeway and on the landward side by triple walls nearly four miles long. The low outer wall provided a cover for archers whose fire commanded a moat 60 feet wide and 20 feet deep, the second wall was 27 feet high with nearly 100 projecting towers and the inner wall was 30 feet high and very thick, with 96 more towers so placed as to cover the area between the towers of the middle wall.

Although challenged by the ever increasing sea power of the Saracens, the Byzantine fleet was always able to maintain communications with the Black Sea. The Byzantine fleet possessed a frightening weapon in its ability to project 'Greek fire' from the bows of its ships, in the manner of a modern flame-thrower. A closely kept secret, the actual composition of Greek fire remains unknown to this day but it is thought to have been a mixture of naphtha, pitch, sulphur, turpentine and quick-lime, the materials being ignited by the quick-lime after water was pumped in. Both from psychological and physical aspects, the effect of this flame-throwing, almost impossible to extinguish, must have been great.

The period of Byzantine supremacy ended at the battle of Manzikert in 1071 A.D. when the Seljuk Turks under Alp Arslan, another group of wagon people or Asiatic horsemen, defeated a motley army of mercenaries and allied nations under the Emperor Romanus. The Seljuk Turks, ablaze with religious zeal, were tough hard-riding horsemen from the Steppes of Central Asia who had begun to raid deep into Armenia and Asia Minor, the principal source of Byzantine armies and revenues. The young Emperor Romanus, said to have been betrayed so that part of his army defected, held the Turks for a day of stubborn fighting. At nightfall, with the Byzantine wings driven from the field, the wounded Emperor was captured and the whole of Asia Minor lay open to gradual infiltration by the Turks.

Although the Byzantine Empire lasted until 1453 it never recovered from this decisive defeat and was finally brought to the ground when Constantinople was captured by the Sultan Mohamet II.

It would be wrong to leave Byzantium without mentioning the Varangian Guard, the personal bodyguard of the Emperors of the East, made up mostly from Scandinavians, Norsemen and Englishmen. Harald Hardrada himself served as a captain of the Varangian Guard for ten years and saw fighting with them in Sicily, North Africa, Egypt and Palestine. In the 10th and 11th centuries the Emperor at Constantinople was the highest bidder for the swords of the hardest fighting mercenary warriors and he had no trouble in filling the ranks of a corps, for the rewards were great and the opportunities for fighting excellent.

## STYLE OF FIGHTING

That the Byzantine Empire lasted for so long was because its army was, in its day, the most efficient military body in the world, displaying outstanding discipline, organisation, armament and tactical methods together with a very high state of morale. The Byzantine style of fighting was based upon a defensive/offensive style of tactics, with emphasis upon the offensive, in that it consisted of a series of successive co-ordinated blows. Their normal tactical formation was capable of great variation and many different combinations depending upon circumstances, the specific enemy and the composition of the force that made up the formation. The basic Byzantine tactical formation consisted of:

(1) Two lines (or blocks) of infantry, the Scutati or heavy infantry in the centre with light troops on their flanks.

(2) Two groups of cavalry behind each flank.

(3) A group of cavalry on each flank of the infantry formations.

(4) Screening units of cavalry thrown out in front and on the flanks.

It was not unusual for Byzantine armies to be composed entirely of cavalry or with only a small infantry contingent—in the latter case, the basic formation would use the infantry in the centre of its second line or else placed as an additional reserve behind the two centre lines which would now be formed of cavalry. The basic formation outlined above assumes that the two arms are combined in about equal proportions, with both infantry and cavalry divided equally between heavy and light.

In this case, the Byzantine infantry would await enemy attack. Should the enemy be cavalry then the infantry were equally confident of their ability to hold them because, with their flanks and rear protected by cavalry, they had all the effectiveness of the Roman legionary against horsemen. When the enemy duly obliged by attacking, the two flanking forces of cavalry would strike him in the flanks, a blow that would shortly be reinforced by the distant screening units coming in on the enemy's flanks and rear. If the Byzantine front lines were overcome, the infantry fell back through intervals left specially for that purpose in the second line—at the same time the enveloping cavalry units would withdraw to regroup and return to the attack. If the second Byzantine line was overcome before the first line had time to rally then the fresh reserve units would be thrown in as a co-ordinated counter-attack, usually in the form of a double envelopment from the flanks rather than as a direct frontal attack.

The outstanding aspects of this standardised tactical system should be noted:

(a) The emphasis on co-ordinated action (missiles/shock and infantry/cavalry).

(b) The emphasis on envelopment rather than direct frontal assault.

(c) The retention (and practical use if required) of a fresh reserve.

In that formation the Scutati (heavy infantry) acted in a defensive role, but whenever opposed by infantry the Scutati (closely supported by missile troops) would seize the initiative and attack. Each tagma (or numerus) was trained to manoeuvre in the same manner as the old Roman cohort and in ranks 16 deep, they attacked like the Romans while hurling their spears just before contact. Byzantine cavalry usually formed in lines 8 to 10 horsemen deep, a formation that engendered confidence in the horsemen at the expense of some mobility.

The Byzantine heavy cavalry, the cataphracti were true descendants of the Roman legionary and the most reliable soldier of the Middle Ages, combining fire-power, discipline, mobility, and shock-action capabilities.

The Byzantine Empire had a number of outstanding leaders, those deserving of an 'Above-average' grading are Belisarius; Narses and Heraclius.

At various times in their history, particularly at the beginning and towards the end, the Byzantine Empire used mercenaries, usually raised from within the confines of their borders. The best came from Armenia, Cappadocia and Thrace whilst the Greeks were considered to be the least suitable military material. The Isaurians were perhaps the most favoured and numerous of the mercenaries employed to fight alongside the soldiers of the Byzantine Empire. Possibly they were not mercenaries in the proper sense in that they could be classed as normal recruits serving for pay as did all Byzantine soldiers. Mercenary cavalry were recruited from the Franks and Turks and, before that, Huns.

For further details of the military system of the Byzantine Empire see Dupuy, pages 214–22.

*Fighting Assessment Chart:* BYZANTINES

| | *Factor* | *Rating* | *Points* |
|---|---|---|---|
| 1. | Ability to assault in controlled order | First Class | 3 |
| 2. | Bravery | First Class | 3 |
| 3. | Control | First Class | 3 |
| 4. | Defensive ability | First Class | 3 |
| 5. | Desperation | Not applicable | – |
| 6. | Discipline | First Class | 3 |
| 7. | Ferocity | Not applicable | – |
| 8. | Manoeuvrability | First Class | 3 |
| 9. | Mobility | First Class | 3 |
| 10. | Movement, speed of | First Class | 3 |
| 11. | Order | First Class | 3 |
| 12. | Tactical Sense | First Class | 3 |
| 13. | Training and Drilling | First Class | 3 |
| 14. | United action, capability for | First Class | 3 |
| 15. | Weapon Handling | Reasonable | 2 |
| 16. | Demoralisation, likelihood of | Unlikely | 3 |
| 17. | Impetuous Action, likelihood of | Rare | 2 |
| 18. | Panic, likelihood of | Unlikely | 3 |
| 19. | Reluctance to take definite action | Unlikely | 3 |
| 20. | Unscheduled Manoeuvres, likelihood of | Rare | 2 |
| | | *Fighting Assessment* | 51 |

# THE CHINESE

A naturally unwarlike people, the Chinese waged war reluctantly under the compulsion of historical circumstances, making no major contribution to the art of war, with the possible exception of a Military Treatise *The Art of War* written in about 500 B.C. Sun Tzu, its author, saw war very much as a struggle between commanders, 'the clever combatant imposes his will on the enemy, but does not allow the enemy's will to be imposed on him'. The early history of China is a long story of civil war and rebellion which weakened the State in its perpetual struggle to protect its frontiers against nomadic tribes invading from the north and west. Nevertheless, these centuries of frontier fighting caused inevitable developments in military organisation and produced a certain military skill.

The very early Chinese armies of about 1000 B.C. were known to have had large numbers of light 4-horse chariots bearing a driver, a spearman and an archer and accompanied by a contingent of about 100 lightly armed infantry. Their style of fighting resembled the earliest Greek warfare as described by Homer. It was a feudal society in which the lords occupied their leisure by fighting for amusement, honour and prestige, riding a chariot into battle in leather armour. During the succeeding centuries warfare became more serious and, although chariots remained the most powerful striking force, large numbers of peasant infantry armed with javelins, swords and bows and arrows were tough enough to form a formidable force. The peoples of the north-western frontier had learned lessons from the Nomads who threatened their borders, so that in time their chariots gradually disappeared and cavalry in the Nomad style appeared in Chinese warfare. In about 400 B.C. the Northern States, frequently invaded by fast-moving horsemen from the Steppes, formed their own cavalry regiments but the expense of equipping such units led to them falling out of favour. The Emperor Wu (140–87 B.C.) incorporated cavalry units in the regular army, considered to be the only arm that could successfully contain the mounted invaders. From this time on, cavalry replaced chariots in the Chinese armies.

In the years 246–10 B.C. was built the Great Wall of China which runs in parts over mountains and gorges for 1,600 miles along the Nomad frontiers south of the Gobi Desert. It was 25 feet wide at the base and 17 feet wide at the top, broad enough for 6 horsemen to ride abreast; 25–30 feet high with crenellated parapets standing 5 feet above the walk and with regular towers. Built at first of rammed earth it was later faced with stone.

Until about 600 A.D. China was a feudal country, racked with incessant civil war between states and walled cities, and constantly under threat from the Nomad hordes whose sporadic raids were held back by the Great Wall. Up to and including this period, Chinese soldiers wore coats and leggings over Lamellar armour. Then brigandine armour came into use, its metal plates being covered with a padded material (silk was often used for high ranking officers) which formed as a lining, a long ankle-length top-coat sometimes with shoulder protection pieces and a short armoured apron at the front. It was worn over high boots and topped by a helmet with a large flapping cloth or brigandine earpiece and neck guard. The plates were secured to the outer padding by metal studs, a distinctive feature of the armoured Chinese warriors. Characteristically, the Chinese liked their armour to be highly decorated, usually red or black or blue for men of high rank—yellow was the equivalent to the Roman purple and worn only by the Emperor. Shields were carried, rectangular for infantry

Ancient Egyptians

Assyrians

Persian Immortals *(By courtesy of Bob Douglas)*

A Greek Phalanx
*(By courtesy of John Chapman)*

War elephants *(By courtesy of Dick Higgs and John Chapman)*

Picts

A brave and lonely Frank holds off the Later-Roman Army

The Vikings land

Cataphracti in the snow *(By courtesy of Bob Douglas)*

Saracen
Horse Archers

Landsknechte and Swiss Pikemen face Janissaries and Spahis, with Syrian Camel-lancer

The Bowmen of England

and circular for cavalry and were highly lacquered and decorated. The uniform was made even more colourful by an exotic array of plumes, pennants and standards of many shapes and sizes. For cavalry, the long jacket was divided at back and front so that the man could sit in the saddle.

By the 7th century A.D. the army was well organised, almost Roman-like. Most of its soldiers were infantry but there were also cavalry and specialist arms such as engineers, etc. At this time, the Chinese, like the Romans, were utilsing the services of barbarians in the ranks of their army, some of them even being allowed to rise to authoritative positions. The professional calling of soldiering produced competent generals who were nominated for campaigns by the Government to whom they had to answer for their actions.

Traditionally noted for their ingenuity, the Chinese are said to have invented the crossbow as early as the 4th century B.C. and seem to have had gunpowder by the 11th century A.D. The first recorded use of gunpowder was during the civil war at the siege of K'aifeng in 1126, but although it was used in mines to destroy walls and perhaps as an explosive missile thrown by a mangonel it is not unreasonable to claim that the Chinese failed to develop gunpowder as the truly lethal weapon it really was.

The Huns who lived in what is today Mongolia were constantly making raids and invasions on Chinese territory so that conflict between them continued for some 600 to 800 years. This long enmity between the nomadic Mongols outside the Great Wall of China and the civilised Chinese who sheltered behind it came to a head when those Nomads were welded into a united force by Genghis Khan and thrown against the mighty Chin Empire. At this time the Chinese depended on powerful fortifications for their defence and their armies were composed of vast numbers of foot soldiers. The first Mongol invasion in 1207 saw them checked by the fortified cities although the Mongol horsemen carried all before them in the field. Profiting from the experience, Genghis Khan trained a number of his officers in siege warfare so that they were able, in 1211, to penetrate the Great Wall and before long the standard with the nine white yak tails stood within the Wall, which had been built to keep out small parties of marauders not an invading army under such a leader as Genghis Khan. For five years, from 1211–1216, the Mongols invaded China in springtime, swept through the country in separate columns before withdrawing in the autumn. These were merely raids on a huge scale, defeating armies and causing widespread destruction but leaving the Great Wall cities alone. As the Mongols gained experience and made intelligent use of captured or renegade soldiers and engineers from Cathay, then they successfully laid siege to many cities. Penetrating the Great Wall in three places, the Mongol horde descended on Peking (then called Yen-King and later Khanblaik). After a long and obstinate siege, the city fell to Genghis Khan in 1214.

At last their repeated onslaughts caused the weak occupant of the throne of the Golden Emperor to flee in 1214. In the confusion that followed, the Mongols moved into Cathay in force and the great Chin Empire collapsed in blood and fire. Fortunately for the Cathayans, Ye Liu Chutsai, a courageous and wise man, whilst a captive of the Mongols, impressed Genghis Khan sufficiently to gain influence over the Mongol leader so saving the lives of millions of his countrymen. He was able to preserve the remnants of the Chin Empire and to establish a government for the newly acquired territory. Kublai-Khan, grandson of Genghis, ruled his vast Empire, embracing all China, Korea, Mongolia, Tibet and much of Siberia, with a perhaps despotic wisdom that he had learned from his wise oriental. Eventually the Mongols overran most of north China, taking Honan in 1233 and thus ending the Kin Tatar dynasty. Southern China was more difficult for Mongol invasion because it was a hilly, wooded terrain unsuitable for cavalry. Gradually extending their rule over south China, the Mongols under Kublai Khan, grandson of the great conqueror, unseated the House of Sung and, after defeating the southern Chinese forces in 1279 in a sea battle off what is now Hong Kong, the Mongols founded the Yuan dynasty in 1280, with the capital at Peking. This dynasty ruled China until replaced by the Mings in 1368.

From China the Mongols absorbed much of military value, for the Chinese, although not warlike, emulated the Byzantines and delighted in warfare as a form of science of tactics, manoeuvres and siege-craft, displaying a passion for the art and science of war, involving to a high point of intellectual refinement elaborate tactics, siege-craft and the use of war engines. Their besetting weakness was a predilection for the defensive, personified by the building of the Great Wall. Chinese craft added to Mongol vigour resulted in a war machine such as the world had never seen up till that time.

The years of Mongol subjection were known as the Yuan dynasty and lasted from 1260–1368, gradually disintegrating into anarchy in the later years of that period. In 1356, Chu Yuan-chang a warrior-monk organised a rebel army that stormed Nanking and began to steadily push the Mongols out of China. Taking the name Hung Wo, this leader founded the Ming dynasty which lasted until 1644 when it was overthrown by a Chinese pirate-general Li Tzu-ch'eng who seized Peking, the capital, causing the Ming emperor to hang himself. The Ming general Wu San-Kuei called in the Manchu army from Mukden and the pirates were defeated. The youthful Shun Chih was set up by the Manchus as the first ruler of the Ch'ing dynasty who ruled China until 1912.

Alarmed and horrified by the superior military technology of Europe that they encountered in the 16th century, the Chinese forsook the art of warfare, retaining a relatively primitive attitude toward military matters until they were forced to Westernise their style of warfare in the 19th century.

## STYLE OF FIGHTING

Although by about 600 B.C. the art of warfare in China was apparently as well advanced as that of the Middle East, it got off to a slower start. History relates that the Chinese, from the earliest times, relied on the bow as their principal hand-weapon so it is reasonable to assume that a large percentage of their infantry were archers although later, because of the lessons learned from Mongol invaders, they became mounted bowmen. As cavalry were not used in the earliest times, the missile effect would not have been backed by the shock effect of cavalry charges so it would appear that, after softening up the enemy, the usual sprawling hand-to-hand mêlée of jostling ill-trained foot soldiers took place. However, from the earliest times, chariots were much used, at first by the leaders and as mobile command posts, but later advantage was taken of the shock action capabilities of the chariot and it became the principal instrument of war, therefore early Chinese warfare can be said to be based upon fire effect from foot archers followed by shock effect by chariots.

Before the 3rd century B.C., the Chinese did not seem to display more than a modicum of control, discipline or tactical sense and their armies probably massed in dense and orderless formations until clashing in the usual sprawling mêlée. Nobles and high-born lords fought from chariots and the 'Hero' or 'Champion' system seems to have prevailed. From now on a gradual change occurred in that warfare became less violent and more ritualistic. This would seem to mean that battles became formalised or stylised in the same manner as those of the Japanese so indicating that set formations manoeuvred and acted in a prescribed, almost ceremonial fashion. With 'chess-like rules', this might well be an interesting innovation for the wargamer looking for something different.

Acting like a feudal society a thousand years before the same situation prevailed in Europe, the Chinese army probably reacted in much the same way by impetuous charges of the knights and ignominious roles for the peasant-infantry.

Later the Chinese, learning from the early Mongol invasions, decided to make cavalry their major arm and it would seem certain that they fought in the usual Asiatic manner. By this is meant that their mounted archers wheeled and swooped in and out as they poured their arrows into the enemy, softening them up for the inevitable charge of heavier cavalry, probably bearing lances. If the enemy were another cavalry army then it was a battle of manoeuvre, won by the best controlled and most mobile of the two combatants. At one stage of history, the Chinese together with the Romans were perhaps the only people who had a regular army organisation that enabled light cavalry to conduct sustained operations—which are largely dependent upon an adequate and efficient forage supply system. It was a period for China, as for most of the other military nations of the world when infantry were completely subsidiary to cavalry. With such a large reservoir of manpower, no doubt Chinese infantry in very large and ponderous formations were used to provide a base of manoeuvre for mobile cavalry action. It is said that by the 5th and 6th centuries A.D. the Chinese, together with the East Romans, were the only military groups capable of efficiently co-ordinating missile power and shock effect brought to a high peak of perfection by discipline, training, leadership and controlled mobility.

At the beginning of the 7th century under a reputed military genius, the Emperor T'ai Tsung, China emerged from four dismal centuries of chaos to enter an outstanding period of external conquest. The Chinese armies reorganised on the lines of the armies of the Eastern Turks so that they were primarily cavalry, eventually defeated the Turks by bettering their own methods.

*Fighting Assessment Chart:* CHINESE

| *Factor* | *Rating* | *Points* |
|---|---|---|
| 1. Ability to assault in controlled order | Reasonable | 2 |
| 2. Bravery | Reasonable | 2 |
| 3. Control | Average | 1 |
| 4. Defensive ability | Average | 1 |
| 5. Desperation | Not applicable | – |
| 6. Discipline | Average | 1 |
| 7. Ferocity | Not applicable | – |
| 8. Manoeuvrability | Average | 1 |
| 9. Mobility | Average | 1 |
| 10. Movement, speed of | Average | 1 |
| 11. Order | Average | 1 |
| 12. Tactical Sense | Reasonable | 2 |
| 13. Training and Drilling | Reasonable | 2 |
| 14. United action, capability for | Average | 1 |
| 15. Weapon Handling | Reasonable | 2 |
| 16. Demoralisation, likelihood of | Average | 1 |
| 17. Impetuous Action, likelihood of | Unlikely | 3 |
| 18. Panic, likelihood of | Average | 1 |
| 19. Reluctance to take definite action | Likely | 0 |
| 20. Unscheduled Manoeuvres, likelihood of | Likely | 0 |
| | *Fighting Assessment* | 23 |

# THE FRANKS

The Franks, a predominantly Germanic people, were fierce warriors with blond or red hair and flowing moustaches. In their early days they wore no armour, did not use a sling or bow and had no horses. They used the sword, dagger and lance (framea) but their great weapon was the Francisca or carefully weighted and balanced throwing axe, which they hurled at the enemy with deadly accuracy. It was from this weapon that they got their name. They attacked with desperate courage in wild rushes after throwing a cloud of javelins at the enemy.

In later years, the Byzantines, far in advance of their time with an intelligence system reporting on the fighting habits and tactics of their enemy, wrote that the Franks believed any form of retreat to be disgraceful—'. . . hence he will fight whenever you choose to offer him battle'. However the Frankish cavalry 'with their long lances and large shields, charged with tremendous impetus. So they should be lured into the hills if possible where cavalry is less efficient.' Campaigns against the Franks should be protracted as much as possible as 'after a few weeks without a battle his troops . . . will grow tired of the war and ride home in great numbers'. The carelessness of the Franks in outpost duties was noted and their indiscipline, so that pretended flight and then turning on the disordered ranks was considered a useful tactic against them.*

The Franks often demonstrated moderation, a very rare quality amongst warriors. In 496 A.D., Clovis intended to slaughter the defeated Alamannis to the last man, but restrained himself when Theodoric the Ostrogoth wrote to him: 'Heed the advice of a chieftain who has great experience in these matters; those of my wars that have had a fortunate outcome are those which I conducted with moderation.' The great barbarian spoke truly—he wisely forbade the pillaging of countries and the massacre of the women and children during attacks on towns.

In 486 A.D., five years after succeeding his father as King of the Salian Franks, Clovis marched southwards from his capital of Tournai (now in Belgium), until he encountered an army under Syagrius, the last Roman Governor of Gaul, at Soissons on the Aisne River. The fierce, axe-throwing Franks overwhelmed the opposition and routed them, killing Syagrius. Clovis then took over all of Gaul between the Somme and the Loire Rivers to firmly establish the Merovingian dynasty. While Clovis was attempting to unite all the Franks under one rule, his reign was threatened by an invasion from what is now Germany when the savage Alamanni tribe swarmed across the Rhine towards Lorraine. In 496 Clovis marched eastwards to meet the invaders at Tolbiacum, near Cologne. In the fierce struggle that followed, the Alamanni at first pushed back the axe-wielding Franks, then Clovis pressed another charge that routed the invaders, driving them back across the Rhine for the last time. Clovis and several thousands of his soldiers were baptised on Christmas Day 496 by the Bishop of Rheims.

After safeguarding his northern frontier against the Alamanni, the energetic Frankish king turned on the Visigoths in the south. He crossed the Loire, and in 507 A.D., at Vouille, a few miles north-west of Poitiers, encountered the army of Alaric. The Frankish infantry overpowered the Visigoths, Alaric was killed and the Goths abandoned their capital at Toulouse and retreated across the Pyrenees to Toledo in modern Spain. Clovis, warmly

* These reports dealt with Western knights, all of whom were called 'Franks' by the Byzantines, i.e. the Westerners showing the usual faults of knightly cavalry.

supported by the Roman Catholic church, now stood supreme as far south as Bordeaux and Toulouse.

In 553 A.D., the last glory of the Eastern Roman army in Italy shone against the barbarian Franks, when 30,000 Frankish foot soldiers, under Buccelin, pushed southwards to Casilinum in Campania. Their method of attack was to advance in a solid wedge, too strong to be broken by a flank attack yet sufficiently mobile to swing round and face the enemy on any side. Narses, successor to Belisarius, arranged his foot archers and dismounted non-Roman cavalry in the centre, flinging out his Roman horse archers in two long wings, one on either side of the main body. As Narses had expected, the Franks came on in a blind, bull-like rush and easily scattered the centre, then they halted bewildered, wondering where next to thrust. Like great gates closing, the Romans swung inwards as though they meant to charge on either flank, prompting the Franks to halt and swing to right and left, braced for the charge. When they came within 100 yards of the Franks, the Roman horse stopped and each cavalryman-archer sat and steadily emptied his quiver into the stationary mass. The Franks were compelled to stand firm under this fire, if they had broken formation the waiting horsemen would then have continued their charge and scattered them. Stubbornly they held for hour after hour while the mounted bowmen picked them off until at last they began to scatter and retreat. Then Narses sounded the full charge and the Roman cavalry, now turning back to sword and lance, rode through and through the scattered troops until Buccelin's forces were utterly destroyed. It was reported that only 5 of his 30,000 men left that field alive.

After Clovis was converted to Christianity in 496, the erstwhile barbarians became the champions of Christendom, to become the most resolute defenders of the Cross for three centuries. They fought with a long iron javelin, its blade shaped like a laurel leaf, and an axe. As they leapt upon the enemy they whirled a small round shield. At various times, they battled against the pagan Saxons of the north, the Lombards of Italy and the Arabs of Spain. Sidonius Apollinars wrote of them: 'From childhood, war is their passion. They are overcome only by sheer weight of numbers or because of a hopeless situation, and if death should bring them low, it is not through fear.'

After the death of Clovis in 511, there was civil strife among his Merovingian descendants and two factions, the Austrasians and the Neustrians, fought a battle in 687 at Tertry in the Somme part of northern France. The Austrasian victory in this apparently minor battle had more important consequences than the usual conflict amongst the Franks because their leader, Pepin of Herstal, provided a unifying influence that he was able to hand down to his son Charles Martel.

In the early 8th century, the Moslems had overrun most of Spain and next rode across the Pyrenees into southern France, taking Toulouse in 718 and Narbonne a year later. In 721 Duke Eudes (Odo) of Aquitaine rallied the Franks and challenged the invaders near Toulouse. The Moslem horsemen were severely beaten by the Frankish infantry and their leader al-Samh ibn-Malik was killed. Recovering, a Moslem-Moorish cavalry horde crossed the western Pyrenees in the summer of 732. At the Garonne River, Duke Eudes (Odo) of Aquitaine, tried to check the heathen flood but was easily brushed aside and the Moslem army of 60,000 hard-riding cavalrymen pressed on towards the Loire. Fortunately for the Franks, Charles Martel (The Hammer) had continued the unifying programme of his father Pepin. He deployed his Frankish infantry in a solid square between Tours and Poitiers, directly in the path of the oncoming Moslem cavalry. The invaders rushed forward in a wild slashing attack but the Frankish infantry held their ground against the mounted attack. In a battle that is said to have lasted some days, the Moslems were beaten but allowed to withdraw unhindered because Charles Martel did not trust his undisciplined Franks to pursue. This sharp set-back, the death of their leader and the frightful casualties they had suffered combined to mark Tours as the high watermark of the Moslem invasion of Western Europe. The victory of Charles Martel is often called one of the most decisive battles of history although the Moslems seemed to have been intent upon plunder rather than conquest. Nevertheless Tours did stop the Saracen advance and help establish the Franks as overlords in France.

At Tours in 732, and for centuries earlier, the Franks and their Germanic kinsmen the Saxons, unlike most other nations, depended upon infantry rather than cavalry in battle, though by the time of Charlemagne even the Franks had taken to the horse. Later in the 8th century, the Frankish cavalryman had become a fully equipped knight, wearing a pliant mail shirt with iron rings; and a round topped and crested helmet with a backpiece to protect the nape of the neck and a round shield, replaced in the 10th century by a kite-shaped shield more suitable to the horseman.

The Lombard kingdom that had become established in northern Italy in 572 stood at odds with the papacy in Rome. Under King Aistulf, in 751 the Lombards seized Ravenna, an ancient city that was technically the Italian capital of the Byzantine Empire, but after years of neglect by Constantinople it was now sought by both Lombardy and Rome. Alarmed at the growing power of Aistulf, Pope Stephen II asked aid from Pepin III (The Short), son of Charles Martel and father of Charlemagne. In 756 Pepin marched across the Alps and with the minimum of military action chased the Lombards out of Ravenna, giving the Pope the conquered territory. This is said to be the foundation of the Papal State and the tacit recognition of the papacy as the temporal successor to the Roman Empire in Italy.

Born in 742, Charlemagne's true greatness lay in the fact that his ideas helped to bring about the downfall of the Dark Ages in Europe and formed the foundation of modern history. He was of pure German origin, spoke German, and for the greater part lived in Germany where he held his court at Aachen in the Frankish Rhineland. He was a tall flaxen-haired man with a long nose and a high-pitched voice, a domed head, a bull neck and a prominent belly and he was never without his golden-hilted sword. A man of immense energy in war, he carefully planned his campaigns. Not assuming the control of a ready-made social and military system, he had the task of creating, against opposition from many directions, a force which could hold in check the aspirations of Byzantium and ultimately lay the foundations of that western feudal system of government which would enable great and unified armies of European knights and soldiers to move against Islam. He believed that the spread of Christianity, by the sword if necessary, was the best means for securing peace and order in Europe. Although the Frankish Empire which he built up fell when other rulers and peoples rose, Charlemagne's concepts of government, law and administration have had a permanent affect on Europe, since they created a form of political monarchy, the effects of which have been seen almost to our own times.

Despite the uncertain temper and heathen beliefs of his barbaric Frankish ancestors, in time Charlemagne was regarded as the champion of Christendom and his wars against Saracen, Saxon, Avar, Slav and Lombard were regarded as Holy Wars and not those of kingly aggression. Through his almost perpetual wars Charlemagne was able to exploit the new unity and superiority of the Franks against allcomers and wherever he conquered he Christianised until the rise of the Frankish Empire in the West became associated in men's minds with the revival of the Church in that part of the world. Sometimes Charlemagne's warlike conversions were not permanent and often he had to retrace his steps to forcibly remind transgressing barbaric tribes of their vows. He had a great deal of trouble with the Saxons who inhabited the marsh and forests in the low lying country about the mouths of the Rivers Ems, Weser and Elbe, a territory which had proved too difficult for the Romans themselves in the days of Augustine and Tiberius. Charlemagne also found them a very tough proposition.

They accepted Christianity in 776 but while Charlemagne was campaigning in Spain they began their pagan practices once more. Returning, Charlemagne defeated them into submission, putting the Cross into their reluctant hands, only for them to again trample it underfoot three years later. This time Charlemagne forcibly transplanted 10,000 families to France and butchered many other thousands in cold blood. A militant missionary, Charlemagne sent out a Frank army to capture the enemy chieftain Wittekind. Locating a large body of Saxons on a ridge north of the Weser River in modern Hanover, the Franks made a brave but foolish charge up the slope and were cut to pieces by the fierce-fighting Saxons. Taking personal command, Charlemagne ranged up and down the Weser Valley seeking Wittekind, and at Verden, on the Aller River, he had 4,000 Saxon prisoners executed when they refused to reveal their leader's hiding place. Charlemagne's 13-year struggle to establish Frankish rule over the Saxons ended in 785 when Wittekind surrendered himself for baptism and submission at Paderborn.

The soldiers of Charlemagne were better equipped than their predecessors, going into battle in armour with lances, swords, daggers and stout shields. The infantry became transformed from a mob of peasants armed with forks, pikes, or axes to an army bearing spears and swords which the nobles, proprietors of the great estates, provided for the State. In his 53 campaigns, Charlemagne took with him siege-machines, provisions for three months, arms and equipment for six months. He established fortified posts, called 'burgs', ringed with palisades, to serve as staging-posts for the flying columns of his armoured cavalry and to defend the ill-defined frontiers of his empire, Charlemagne had fortresses or strongpoints built. They usually consisted of a round wooden tower built on a mound surrounded by a wooden palisade with a deep trench on either side of the fence. It was from these towers that the moated castles of the Middle Ages were later developed.

Five years after becoming King of all the Franks, Charlemagne marched into Italy to aid Pope Adrian I against the Lombards. Crossing the Graian in two columns, Charlemagne outflanked the Lombard army of King Desiderius at the Mount Cenis Pass. The Frankish cavalry swept down the Po Valley to Pavia,* where the Lombard King took refuge behind the strong walls of the city in September 773. Lacking siege engines on this occasion, Charlemagne deployed his troops round the city and settled down to starve out the enemy. Finally in June 774, Desiderius surrendered his city and kingdom and Charlemagne proclaimed himself King of the Lombards. The Frankish king made a quick return trip to northern Italy during the winter of 775–6 to quell a rebellion in Treviso—thereafter the Lombard kingdom vanished from history.

In spring 778 Charlemagne led an army of Franks against the Moslems. Crossing the Pyrenees, he entered Pamplona and then pushed down the Ebro to Saragossa, a city of high walls defended by a Moslem army under Abd-er-Rahman, but his siege was interrupted by the need to put down a Saxon revolt east of the Rhine. Crossing the Pyrenees on the return to Spain, the baggage train and rearguard were ambushed by mountain Basques in the Pass of Roncesvalles. In a battle notable for its inspiration of medieval literature, not a Frank escaped alive and among the dead were the paladin Roland, Prefect of the Breton March.

The barbarian Avars had long worried Charlemagne with their threat to eastern Bavaria and north-eastern Italy, and a Frankish expedition along the Danube in 790 had failed to subdue this fierce, elusive tribe. In 795, Charlemagne sent out one expedition from Bavaria and another from Italy, to join forces beyond the upper Danube. Forcing their way across the Tisa River, a northern tributary of the Danube, the Franks reached the Avar stronghold, a settlement enclosed by a huge ring of earthworks. Eric of Friuli led the Frankish charge that routed the barbarian defenders and the Avars vanished from history, after being offered the choice of baptism into the Christian faith or death.

Crowned Holy Roman Emperor by Pope Leo III on Christmas Day 800, Charlemagne spent the remaining years of his life trying

* The Monk of St Gal wrote of this:

'Then appeared the Iron King, crowned with his iron helm, with sleeves of iron mail on his arms, his broad breasts protected by an iron byrnie, and an iron lance on his left hand, his right free to grasp his unconquered sword. His thighs were guarded with iron mail, though other men are wont to leave them unprotected that they may spring the more lightly on their steeds. And his legs, like those of all of his hosts, were protected by iron greaves. His shield was plain iron, without device or colour. And around him and before and behind him rode all his men, armed as nearly like him as they could fashion themselves; so iron filled the fields and the ways and the sun's rays were in every quarter reflected from iron. "Iron, iron everywhere" cried in dismay the terrified citizens of Pavia.'

to keep his large empire at peace. The Northmen were his only persistent remaining enemy but Charlemagne managed to protect his territories and keep the Danish frebooters in check by a powerful North Sea fleet and a chain of fortresses along his northern frontiers together with a series of treaties with the Northern Kings. Also, he fought Arab raiders on the Mediterranean. After the death of Charlemagne in 814 at the age of 71, Frankish domination in the East faded and his Empire decayed under his successors, leaving the Pope without a champion. Charlemagne's successors were unable to hold off the Northmen and by 881 they had struck at the very heart of Charlemagne's Empire, burning even Imperial Aachen. Unable to stand against such attacks the Holy Roman Empire began rapidly to disintegrate so that by the end of the 9th century it consisted of little more than a number of decentralised kingdoms mis-ruled by militant knights.

## STYLE OF FIGHTING

The Franks were essentially foot soldiers. At times they did make some use of cavalry although, in their hey-day, the Franks were perhaps the outstanding exception in a period that placed great reliance upon horsemen. Later Frankish experiences against cavalry armies gave them a healthy respect for horsemen and by the end of the 6th century they too had come to rely upon the shock action of heavy, lance-carrying horsemen.

The Franks assaulted in dense disorganised masses of highly mobile infantry, combining shock effect with missile power as, immediately before contact, they would hurl their heavy throwing axe, their javelin and then take advantage of the resulting confusion to dash in with the sword. They awaited cavalry charges in dense masses which absorbed the shock of the charge so that they could mill around the halted horsemen cutting down both horse and rider.

They owed much to extraordinary physical strength coupled with great courage in a period when their enemies tactical skill had deteriorated.

Clovis, their outstanding leader, had a measure of control over them and they can be said to have been reasonably disciplined.

At the Battle of Tours (732 A.D.) Charles had good tactical control over his force although his heavy cavalry were relatively undisciplined and extremely difficult to control when once loosed, besides being slow-moving in comparison with their lighter Moslem counterparts. For this reason, for this battle Charles dismounted his cavalry and formed them into a solid infantry phalanx positioned on a commanding piece of ground.

Their Moslem opponents were highly effective attackers who never missed the chance of exploiting a gap in a battle line, but if the formation against them was solid and organised they lacked the weight to effectively shatter it by shock action. A favourite Moslem tactic was to lure the cumbersome Frankish cavalry in a pursuit and then, when they were spread out, turn and slaughter them piecemeal.

Before Charlemagne, the Franks were noted for coupling great vigour and courage with a minimum of discipline so that, even under able commanders such as Charles Martel and Pepin, Frankish armies were unreliable. Charlemagne harnessed Frankish vigour in a disciplined and efficient organisation under his strong personal leadership. Charlemagne reintroduced the bow to his armies and evolved a standard tactical system so that his subordinate commanders carried out his orders in the manner he wished.

See Dupuy, pages 208–9, for further details of the military system of Charlemagne.

*Fighting Assessment Chart:* FRANKS

| | *Factor* | *Rating* | *Points* |
|---|---|---|---|
| 1. | Ability to assault in controlled order | Reasonable | 2 |
| 2. | Bravery | First Class | 3 |
| 3. | Control | Average | 1 |
| 4. | Defensive ability | Average | 1 |
| 5. | Desperation | Not applicable | – |
| 6. | Discipline | Average | 1 |
| 7. | Ferocity | Not applicable | – |
| 8. | Manoeuvrability | Reasonable | 2 |
| 9. | Mobility | First Class | 3 |
| 10. | Movement, speed of | Reasonable | 2 |
| 11. | Order | Average | 1 |
| 12. | Tactical Sense | Reasonable | 2 |
| 13. | Training and Drilling | Average | 1 |
| 14. | United action, capability for | Reasonable | 2 |
| 15. | Weapon Handling | First Class | 3 |
| 16. | Demoralisation, likelihood of | Rare | 2 |
| 17. | Impetuous Action, likelihood of | Average | 1 |
| 18. | Panic, likelihood of | Rare | 2 |
| 19. | Reluctance to take definite action | Rare | 2 |
| 20. | Unscheduled Manoeuvres, likelihood of | Average | 1 |
| | | *Fighting Assessment* | 32 |

# THE JAPANESE

The early history of Japan is full of tales and legends of war, bloodshed, treachery and noble deeds. Originally, the Japanese were a Mongol people who made their way from the high plateau of the continent of Asia to the islands they now occupy. They found them inhabited by the Hairy Ainus, a strange and primitive race of hunters and fishermen who so strongly resisted the invasion that hostilities lasted for centuries until about 800 A.D. when Saknaove Tamuramaro subdued the northern island of Hokkaido and was given the title of Sei-i-tai-Shogun or Barbarian-Subduing-Generalissimo by the Mikado.

By the 1st century B.C. there were about 100 tribes or clans in Japan, all fiercely antagonistic towards each other. Their ambitious chieftains began importing iron swords and armour from China and in the 4th century A.D. the people of Kyushu emerged as leaders of the clans to establish a centralised government at Yamato on the main island.

As with all island peoples, strong seafaring instincts and traditions developed over the years, reaching their peak in the period 1900–1945. Throughout Japanese history, numerous expeditions have gone to the Korean peninsular, beginning in 369 when Southern Korea was successfully invaded, and in 391 Japanese forces penetrated as far north as Pyong-Yang.

The next three or four centuries saw order and unity slowly making progress so that a national military system began to take shape with more than 700 specific laws promulgated with the hope of evolving a State army. One such law conscripted the peasants as a national reserve, but the ill-equipped peasants showed such resentment that the scheme was changed to one of each Province maintaining its own force of trained regular soldiers recruited from the upper classes, which stressed the gulf between the peasant and the warrior classes in Japan.

Japan was a feudal society until the 17th century, its Emperor a puppet because power lay in the hands of the feudal lords (the daimyos) and the warrior-clans they led. Lacking stability because of its lack of political unity, civil war frequently raged as clans and their war lords struggled for power. The warriors-clans lived in and around castles of a particular oriental appearance, surmounted by a five- or six-storey stone keep, crowned with typical curved roofs, with surrounding walls and moats set in an irregular and complicated pattern or system. Little is known about the use of siege-engines but it seems extremely likely that the people as ingenious as the Japanese would have devised them although their peculiar power-system built around the proscribed codes of the Samurai-Knight may well have caused them to disdain such mundane weapons of war as mangonels, catapults and siege-towers.

The beginning of the Kamakura feudal period (1185–1333) grew out of a civil war between the two most powerful clans, the Taira and the Minamoto who clashed in a naval battle at Dan no Ura (the eastern end of Shimonoseki) on the south-western end of Honshu in 1185. Led by Yoshitsune, the Minamoto force, totally defeated by the Taira and Antoku, the child-emperor, supported by the Taira, was killed in the battle. Yoshitsune's older brother Yoritomo became the virtual ruler of Japan and his descendants remained in power until replaced by the Hojo family in 1219. The Shogun Yoritomo of the Minamoto regularised the Japanese system of dual authority where the Mikado was the ceremonial head of the country with the actual power in the hands of the Shogun. He set up the Bakafu, a military-type government, with the warrior caste, the Samaurai, controlling the administration and social life of the country. As in any feudal state, there were constant struggles for power among the great nobles and barons with the common people reduced to serfdom. The daimyos lived in great luxury, surrounded by Samurai retainers whom they supported.

The warrior-class had their military cult, compounded of Zen Buddhism, spartan endurance, loyalty and work, with the sword as a symbol of rights and status. They dominated Japanese life. Even monks, like medieval Templars, turned their monasteries into fortresses and waged war with the same horrifying intensity as the Teutonic Order. In the 16th and 17th centuries, this military cult was exalted into Bushido, a national cult which combined the aesthetic and mystical aspects of Zen and Shintoism with the strict military code of the warrior—a combination that was to become one of the chief characteristics of the Japanese people.

In the civil wars of the 12th century it became the custom for the defeated leaders to lose their lives together with those of their families—all wiped out in a merciless and barbarous fashion. As the frequent battles between small clans and factions invariably culminated in the complete wiping out of the defeated clan, in time the knightly clans were faced with extinction. In characteristic Japanese fashion the problem was solved by hara-kiri—defeated leaders killing themselves in a proscribed ritual fashion so absolving their families from blame and sacrifice at the same time as they cleansed their own honour. After a battle, the vanquished knelt down by the hundred and performed this bloody act.

Feudalism in Japan took long to die, the last remnants only being abolished at the Meiji restoration in 1870, with the symbolic end coming five years later, when the wearing of swords was prohibited. In Europe feudalism disappeared far earlier; yet there are interesting comparisons between the knight and the Samurai; Bushido and the Western code of chivalry.

Although there were attempts at military tactics in the form of elementary surprise attacks and ambushes, strategy and tactics played little part in Japanese battles of the 12th and 13th centuries, the only ordered procedure being the ritual before combat could start. With both armies drawn up facing each other, the attacking side fired a 'singing' arrow (a shaft with a whistle-device attached to it) then its commander cried out 'Ei ei!' which was taken up by his men and in turn by the opposing commander and his army. Only then could the two forces come together and commence fighting. There was no tactical manoeuvring or overall control and, as a Samurai won a great and honourable reputation by being the first to ride up and make contact with the enemy, it was likely that each battle developed into a 'free for all' with Samurai-Knights riding forward, leaving their bands of retainers (chugens) to follow up at their own pace and in their own manner. Each Samurai felt free to lead his band in any direction he desired and the band, often left leaderless, could retreat or dally if they felt disinclined to contact the enemy. Thus a battle resembled a gigantic multiple fencing match with each Samurai selecting an individual opponent and fighting him to the death. The Samurai, weighted down with gorgeous armour, gold and silver plated and tied together with many coloured cords of silk or leather, was a master of the bow and the sword—the long sword, sometimes 5 feet long for fighting, wielded with both hands, and the short sword for beheading the fallen foe or for committing suicide (for hara-kiri was a part of his courage). Swords made by the master craftsmen of Japan have seldom been surpassed.

The Samurai Knights, professional soldiers, strongly dedicated to an ascetic and vigorously defined code of rules, were the backbone of every army. 'As among flowers the cherry is queen, so among men the Samurai is lord' was a favourite Japanese saying.

The profession of arms had become an honoured one in Japan. Faithful to their overlord, impassive and indifferent to suffering, fond of the arts (the sword-wielding Samurai delighted in verse, painting, the art of landscaping together with the formal and exquisite tea ceremony in a manner akin to the cultural accomplishments of some Western knights of the later Middle Ages), the Samurai lived to a strict code, killing others as readily as they commited hara-kiri. Clad in a coat of mail, a cloak, wearing plaited straw sandals and not only a helmet but also a mask, this was the feudal warrior with his two swords who still existed in the middle of the 19th century. The peasants and lower classes were subservient to the Samurai who, like the feudal knight of the West, was above the law; the relationship being similar to that of a Saxon churl and a Norman knight. Yet the code of a Samurai-Knight made it unthinkable for him to desert his chugen, who were sometimes more of a liability than an asset, often being men of poor fighting qualities and low morale probably only stirred to fight desperately when their lives were at stake. A Samurai-Knight had a varying number of chugens, who served him, fighting on foot and forming a loose fighting unit. They wore a light armour in the shape of helmets and corselets and were armed with spears and short swords. The poorer Samurai-Knights were known to serve as chugen.

The protective armour of the Samurai was a garment of iron and leather held together with the silk or leather cords, and a horned metal helmet. Some armour was decoratively inlaid with precious metal. Grotesque and unwieldly as Japanese armour may appear to Western eyes, it was actually light and flexible, well adapted to the Japanese style of fighting. It has been described by a leading authority as a 'deftly woven complex of steel laminae, leather, silk, cotton padding, each in a special defence for a special region of the body'. In beauty of design and excution of the individual parts, the best pieces of Japanese arms and armour are unequalled. Even the lowly arrow head was often enriched with fretwork and engraving, whilst such objects as the hilts and sheaths of swords and daggers were often masterpieces of the metalsmiths' and lacquer workers' art. The chief weapon of Samurai was a bow, about 6½ feet in length, drawn well behind the ear and resembling the English longbow both in appearance and effectiveness.

The Japanese never had a proper war horse but rode to battle on small sturdy ponies, rough and undignified in appearance but possessing qualities of endurance and courage. Sometimes they were armoured, but not until the 16th century.

A feature of battles in this period was the Samurai-Knight's challenge to individual combat, when he stepped from the ranks and recounted a saga of his noble ancestry and their valiant deeds, concluding with an embellished account of his own feats of arms. His opponent similarly offered his own 'commercial' and then both knights clashed, fighting in a bizarre and stylised manner with their long swords. The Samurai fought each other so valiantly that each commanded the respect of the other but every such duel had to end with death to one of the fighters. Though the winner might wish to spare his adversary, honour demanded that he must kill him and so the loser was beheaded. Loyalty, simplicity, devotion, honour and self-control was the code of the Japanese warrior, bringing about a native tradition, a glorification of the individual warrior dating from very early times. Isolation from the rest of the world perpetuated and prolonged this feudal idea of the individual warrior and it was not until the close of the 17th century that the Japanese began to realise the value of concentrated attacking movements.

In 1274 a Mongol–Korean fleet crossed the sea of Japan, seized the Islands of Tsushima and Iki and then made a landing at Hakata Bay in northern Kyushu. Mongol weapons and organisation proved far superior to those of the Japanese, and the custom for a single Samurai to advance like a Homeric hero to challenge the enemy to combat was countered by the Mongols not allowing a single warrior to step forward in the prescribed Japanese fashion. Instead the wily and disciplined Mongol ranks would open up to let the Samurai-Knight pass then slowly close in around him and cut him to pieces. After winning a bridgehead on the island, the invaders returned to Korea because of a storm. The Hojo rulers of Japan, who had succeeded the Minamoto clan, immediately began building fortifications against further invasions. Japan was invaded again in 1281 and, like the first attack, the Mongols were beaten off by a fortuitous combination of bad weather, ill-luck and a policy of desperate self-sacrifice on the part of the Japanese. From these invasions, the Japanese are said to have acquired a knowledge of the value of infantry and tactical manoeuvring in battle.

Learning lessons from the Mongol invasions, the bands of Samurai retainers were replaced by infantry (ashigaiu), said to be reasonably well drilled and disciplined, armed and equipped in a more or less standard manner. They wore lacquered cuirasses, sometimes adorned with the badge of their lord, and light helmets that resembled flattened cones of metal or leather. At first they were armed with bows, spears and swords but later some of them used matchlock muskets that had been brought to Japan by the Portuguese. Many soldiers flew their lord's banner from a light bamboo pole attached to the back of his cuirass. The size of armies varied according to the wealth and power of the clan or its daimyo. Similarly, the number of Samurai-Knights with each force varied in accordance with the ability of the daimyo to pay for their hire. The Samurai became less powerful and no longer fought individually, being grouped into troops of cavalry, wearing the banner of their war lord on their cuirasses.

The monotonous record of civil war throughout the 14th to the 16th centuries has the single interesting development in the emergence of the Ashigaru, a class of peasant soldier, with little or no armour and only a single weapon, perhaps a sword, spear or halberd. They fought in bands with no other allegiances and, because they resembled brigands they acted like them by burning and looting.

Medieval Japan had considerable intercourse with China and Korea and at times interfered with Korean affairs. In 1592 Toyotomi Hideyoshi led a successful invasion of Korea which involved fighting the Chinese—the campaign ended with a Japanese withdrawal. Civil wars continued until Tikugawa Ieyasu took over control of the entire country after the battle of Sekigahara in 1600 and Japan settled down to 200 years of peaceful isolation.

## STYLE OF FIGHTING

For many centuries a sort of 'oriental' feudal system prevailed in Japan, when Samurai-Knights each with their own band of retainers fought their part of the battle in the manner of their own choosing. There was little control or discipline and no tactical planning. The Samurai-Knights often fought as 'Champions' with similar warriors from the opposing side—on some occasions their chugen or band of retainers acted as spectators and on others they milled together in a mob-like mêlée.

Later, the Japanese style of fighting became more formalised as the ceremony and ritual, so dear to this race, began to assert itself. It would seem that, as with the Chinese, their battles took on a stylised form with groups of soldiers performing prescribed manoeuvres. Again, there would appear to be little tactical ability. The Samurai-Knights were highly self-disciplined and expert in the use of their weapons. In the sense that they dominated the Japanese military scene, they were competent and respected leaders although their followers seem to have had little training or to have been handled with very much control.

*Fighting Assessment Chart:* JAPANESE

| | *Factor* | *Rating* | *Points* |
|---|---|---|---|
| 1. | Ability to assault in controlled order | Average | 1 |
| 2. | Bravery | First Class | 3 |
| 3. | Control | Average | 1 |
| 4. | Defensive ability | Average | 1 |
| 5. | Desperation | Not applicable | – |
| 6. | Discipline | Average | 1 |
| 7. | Ferocity | Not applicable | – |
| 8. | Manoeuvrability | Average | 1 |
| 9. | Mobility | Average | 1 |
| 10. | Movement, speed of | Average | 1 |
| 11. | Order | Average | 1 |
| 12. | Tactical Sense | Average | 1 |
| 13. | Training and Drilling | Average | 1 |
| 14. | United action, capability for | Average | 1 |
| 15. | Weapon Handling | Reasonable | 2 |
| 16. | Demoralisation, likelihood of | Average | 1 |
| 17. | Impetuous Action, likelihood of | Likely | 0 |
| 18. | Panic, likelihood of | Average | 1 |
| 19. | Reluctance to take definite action | Average | 1 |
| 20. | Unscheduled Manoeuvres, likelihood of | Likely | 0 |
| | | *Fighting Assessment* | 19 |

# THE INDIANS

From earliest times immigrant peoples crossed the passes of the North-West Frontier—in the years between 2400 B.C. and 1500 A.D., first Greeks and then Turks, Huns, Mongols and Persians pushed southwards. They easily defeated the uncohesive armies of India and it was the Vindhya Range of mountains in Central India that halted most invaders. Usually they came in October or November when it was possible to live off the country; between June and September armies could not move because of the monsoon rains.

The first notable record of Indian warfare was the invasion by Alexander the Great in 327–325 B.C., when he crossed the Hindu Kush, captured the city of Taxila and defeated King Porus at the battle of the Jhelum (Hydaspes). Undoubtedly a leader of courage and possibly the ablest Hindu commander, Porus had little chance against Alexander. His army included elephants and chariots, light two-horse chariots made of wooden struts bound together with thongs and drawn by two horses, crewed by a driver and a bowman. In addition, there were heavier four-horse chariots carrying two shield bearers, two archers and two drivers, who dropped the reins in the mêlée to become javelin men. Unfortunately, the chariots made little impression at the Jhelum, getting stuck in the mud.

Indian warfare had a tradition of elephants, and for many centuries they were the principal offensive power in Indian armies, retained amongst the Hindus and Moslems until their effectiveness was destroyed by the use of firearms in the 17th century. At the Jhelum, the Indian front line was composed of 85 elephants deployed 100 feet apart, the spaces between being filled by picked infantry. Porus himself rode into the battle on an elephant. Each elephant carried a mahout (driver), and three warriors armed with a choice of bows and arrows, javelins, knives, pots of oil or stones. The animals themselves were elaborately equipped with mail armour and decked with necklaces, rugs and heraldic devices. Nevertheless, they made a poor showing in spite of their strength and terrifying appearance that was calculated to strike terror into the heart of all but the best trained soldiers and horses. They could trample men under foot and batter down obstacles but they were too difficult to control to be trustworthy as the main offensive weapon in battle. There are numerous instances, from the Jhelum onwards, of wounded or panic-stricken elephants turning and causing chaos in their own forces. Nevertheless, they were always included in every Indian army—one commander, Chandragupta Maurya (322–198 B.C.), had an elephant corps 9,000 strong.

Apart from the unjustifiable reliance placed on elephants, organisation and composition were the main weaknesses of ancient Indian armies, partly due to a feudal method of recruiting which made difficult unity, command and standardisation of equipment.

In the earliest times it is likely that there was an 'heroic' period in Indian warfare, similar to those in Greek and Japanese warfare; in India ancient battles were often multiple duels between aristocratic charioteers accompanied by retainers on foot. Their wars were petty affairs, often stimulated by politicians for prestige, pride or other obscure motives. Only the most intimate supporters of one side or the other took any real interest in these struggles. All these weaknesses were exposed in later years by the Islamic invaders.

The soldiers forming Indian armies were widely different types ranging from the professional soldiers of the warrior-caste to ill-armed savages. Roughly, they fell into six categories—hereditary troops; mercenaries; levies or municipal guards; feudal-type contingents from allied chieftains; troops captured or seduced from the enemy and men of the jungle tribes. In later years the Hindu army was fundamentally composed of regular corps of professional soldiers maintained by powerful princes and rulers, to whom were allied feudal and other levies raised when the situation demanded. As in most armies of the day, the ill-trained levies were employed on labouring tasks, porterage, foraging and digging fortifications. During the 'Heroic Period' the infantry were probably relegated to lookers-on, but after that they were usefully employed in hilly country or as garrisons and particularly in battles where numbers alone might be decisive. Not rated as highly as chariots or elephants, although rated superior to infantry, cavalry were mounted on ponies because the native breeds of horses in India have always been inferior to those of the West and Central Asia. This was a great weakness in Hindu armies and defeats at the Jhelum and later by the Turks can be attributed to the enemy possessing well mounted cavalry under competent commanders. In time, two types of cavalry emerged—the heavies who were positioned in the centre line and delivered charges and the lights who were positioned on the wings for scouting, flanking, encircling and pursuit. Long lances were used in the charge and swords in the mêlée but mounted archery was never developed by the Hindus. Small and large chariots, with crews of from two to twelve men, remained in service in India until the 8th century A.D. They were more slow and unwieldy war carts than chariots.

Musical instruments were used to inspire the soldiers, to signal orders and to keep them in step when marching.

Early Hindu soldiers used maces and club, the nobles favoured battleaxes but the bow was the most favoured weapon in India and was used from the 4th century B.C. until the 18th century A.D. The earliest bows were wood but composite bows were later used, most were short but details exist of a bow the height of its user that, with the aid of the foot, fired arrows that penetrated both shield and armour. Besides the bow, stones and lead pellets were thrown from slings, whilst other missile-men hurled quoits and a discus with sharpened edges. Hindu soldiers also carried swords and various types of spears (shorter than the European pikes of the 17th century) and javelins. Everyone except archers carried shields made of ox-hide or even tiger-skin stretched over bamboo, some primitive shields were made of matted creeper. They varied widely in shape and size and were decorated with emblems. Body armour, probably chain mail, was worn by the wealthy whilst the lower ranks wore padded coats of quilted cotton.

Elephants, camels, pack-ponies, bullocks and bullock carts formed the transport of a Hindu army making it a spectacular sight on the march, moving sluggishly in a great mass accompanied by music and shouting. Elephants and the entourage of powerful rulers were made colourful by flags, feathers, silks, rugs and parasols with precious metals gleaming in the sun. Every army had a tail made up of vast numbers of camp-followers—traders, musicians, prostitutes, etc., etc.

Hindu soldiers were courageous but their commanders lacked vision and possessed little strategic or tactical sense. If the commander of an army fell, as happened to King Dahir of Sind in 712 A.D., his army was liable to disintegrate and there are many instances in Indian history of wholesale desertion and changing from one side to another.

The earliest tactical formations are difficult to discover because the chroniclers gave them symbolic titles such as 'Hawk'; 'Rohmboid' or 'Crocodile'. However, it seems that there were four basic formations—the 'staff'; the 'snake'; the 'circle' and 'detached order' each with its own variations. Before a battle, much time was usually occupied forming troops into conjectural battle deployments but once the actual fighting began all discipline and tactics were thrown to the winds when turbulent masses of men hurled themselves at each other whilst 'champions' fought to the death in separate 'duels'.

From the 10th to the 7th centuries, ruthless Turko–Islamic tribesmen invaded across the borders into this rich and dis-united country, rarely failing to defeat the Hindu king or prince who stood against them. The first conquest provided the springboard for the next so that Hindu territories were devoured one by one as the forces of Islam progressed south and eastwards. The early invaders settled on the land they had conquered and became Indians, to be swamped themselves by succeeding raids from the north-west in a sequence of conquest that flowed uninterruptedly for eight centuries. Usually, each separate raid was a brief affair but occasionally an outstanding Moslem conquerer swept all before him in a devastating tide. It was not until the 17th century that the tribesmen of the Assam jungles halted the then decadent Moghul forces.

The first and possibly the greatest Turko-Islamic conqueror of India was Sultan Mahmud of Ghazni (997–1030), one of the most aggressive rulers of the day, who is said to have led seventeen campaigns in India in 29 years until his empire extended from

Persia to the Ganges. In the first of these raids in 1001, he took 10,000 Afghan troops through the Khyber Pass into what is now Pakistan, to come upon a force of some 40,000 Punjabis and 300 elephants at Peshawar. They were routed by the hard-riding Afghan cavalry who went on to plunder the region.

The Hindus lacked in outstanding measure those essential military qualities possessed by the Turks who battered down complacency and tolerance with the terrifying fire and vigour expected of fanatical barbarians devoted to the faith of Islam. Highly mobile on speedy, tough Turkoman and Arabian horses, the virile, united Turks displayed a contemptuous disregard for death as they pushed aside the softer Hindus in a series of raids and invasions planned by commanders of talent and sometimes genius. Hordes of mounted archers, in the old and effective tradition of the Parthians, the Huns and the Mongols, formed their armies, and their composite bow was as good as any weapon of the Hindus, besides they used it to better effect. Turkish tactics were to harass and exhaust the enemy by swarms of elusive mounted archers circling with a controlled mobility, and then to punch home a charge of heavier cavalry. The Turkish tactics at Second Tarain (1191) were identical to those used at Manzikert.

The Persian Sultan Mohammed of Ghor (Shihabuddin Ghori) subdued Sind in 1182 and the Punjab in 1187 in the course of a series of invasions. Four years later on the Saraswati River, his greatly outnumbered Turkish army was defeated by the Hindu King, Prithviraj. Characteristically, the Hindus did not energetically follow up their victory in any strategic sense, so permitting the Moslem Sultan to return 18 months later and at Taraori near Delhi, not far from the site of the earlier defeat, his mounted archers crushed a Hindu army. Restricted by disunited command because of caste inhibitions, the Hindus' clumsy tactics brought about their own downfall. Mohammed occupied Delhi and made all north-western India his own.

In the autumn of 1221 Genghis Khan, with 40,000 mounted warriors, was pursuing the Sultan Jelal-ad-din through Bamian Pass in the Hindu Kush Range. The Moslem with some 30,000 men turned to make a stand with his back to the Indus River, right flank protected by a bend in the river and left flank anchored against a mountain ridge. The Sultan opened the battle with a cavalry charge by his right wing, and as this attack gained ground, he sent forward his centre, reinforcing both sections with detachments from his left flank. Just as victory seemed within the grasp of the Moslems, 10,000 Mongols sent by Genghis Khan over the seemingly impassable ridge on the Moslem left, debouched from the heights, while the great Khan led a counter-attack on the other side of the battlefield. The Moslem army was cut to pieces, many drowning in the Indus, Jelal-ad-din escaped by swimming his horse to the far bank but he was now a ruler without a country or an army and the Mongol hordes ruled from Tibet to the Caspian Sea.

After subduing Persia and the Golden Horde, the Mongol conqueror Tamerlane turned to India, marching south-east from his capital of Samarkand, crossing the Indus in September 1398. On December 17th Tamerlane's cavalry host destroyed the Indian army of Mahmud Tughlak at Panipat. Plundering Delhi, the Mongol invaders razed the city to the ground before turning north-east and crossing the Jumna River. Marching to the heavily fortified city of Meerut, the Mongol leader started mining operations against fortifications considered to be impregnable, but on January 9th 1399 his troops, impatient with the tedious siege operations, stormed the city with ladders. The Indian garrison was massacred and the fortifications destroyed by miners.

Tamerlane's sixth descendant, Babur Tiger, descended upon India in 1525. Crossing the Indus River he encountered the Afghan Sultan of Delhi (Ibrahim Lodi) at Panipat, 53 miles north of Delhi, with an army much larger than his own and including elephants. The Mongols completely defeated the Afghan, killing Ibrahim and ending the Lodi dynasty. Subsequently destroying the Hindu–Rajput Confederacy at Sikri near Agra, Babur occupied Delhi and Agra, laying the foundation of the great Mogul dynasty, formed by his grandson Akbar, that remained in power until 1761 and survived in part until 1857. Before his death in 1530, Babur's domains extended from the Oxus to the frontier of Bengal, from the Himalayas to Gwalior.

Settling in India, the Turks lost the barbaric vigour of the Steppes, absorbing the ancient and less virile ways of their new land. Furthermore the advent of firearms affected their warfare—at Panipat, Babur's army was different from those of his predecessors in that he had fewer Turkish horsemen (who were his elite troops) together with contingents of native Indians fighting with traditional bows, swords and spears and infantry armed with matchlocks and falconets mounted on carts. He incorporated elephants in his array but the mobility of his cavalry declined.

Although firearms and artillery were used in India as early as the mid-14th century, together with rockets and incendiary substances, some Indian troops were still effectively using bows and arrows as late as the Mutiny in 1857. Babur skilfully used artillery in the early 16th century and in 1565 at the battle of Talikota, Nusain Nizam Shah, a good Indian general, is reported to have pushed his artillery well forward, screened by horse archers.

As strong as the best in Europe, some remarkable forts existed in India, such as that of Mandu in Gujerat built in 1406–35. Its construction difficulties must have been immense as it was built on irregular ground with scarped sides on a hill 1,000 feet above the plains and crowned by powerful walls, bastions and gates. Other notable fortifications in India are those of Agra, Daulatabad and Madura. Indians used catapults and the usual siege-weapons but sieges were rare as opposing generals generally skirted round and avoided the great strongholds.

## STYLE OF FIGHTING

By 600 B.C. Hindu armies were made up almost entirely of footmen with the bow as their principal weapon and because horses were reserved for pulling the two-man war chariots of the kings and nobles there were no cavalry. Thus it is likely that Hindu warfare was a shapeless affair with the fire effect of the archers being followed by a disorderly mêlée of ill-controlled infantry. During the next two centuries the chariot became the decisive element in Indian warfare and in about the 6th century B.C. war elephants began to appear on Indian battlefields. The normal way of fighting was to try to reach a decision by long-range archery fire. If this was not possible then confused mêlées took place between archers and javelin men armed with swords. It is likely that at this period the Hindus used 'Champions' who fought individual duels with the 'Champions' of the opposing side.

The first great military figure of Indian history was Chandragupta (about 323) who is said to have fought against the Macedonians and to have built up the power of India's first great dynasty, the Maurya Empire. During this period standing armies became normal and although the importance of training and discipline was well known commanders found it difficult to impose such military fundamentals upon their troops in practice. They relied much on elephants and chariots, the former were unreliable and the latter unwieldy in the face of competent trained infantry and cavalry. Nevertheless chariots continued to be used in India for another thousand years whilst elephants were brought into battle right up to the 19th century.

During the last two centuries before the birth of Christ cavalry became the major arm in northern India and the principal weapons the lance and short sword, but in southern India, largely because of the difficulty in raising horses of the required quality, cavalry were less important and horses were used to draw chariots.

For the first two centuries A.D., Hindu warfare was conducted in accordance with formal and elaborate rules and, in the north, was more a sport of kings than a struggle for existence. For that reason battles were conducted with far less savagery than elsewhere in the world and defeated foes were treated with chivalrous generosity. Such attitudes made it easy for barbaric invaders to overthrow the soldiers of the country, but strangely, these invaders appeared to adopt the martial habits of the country as well as its culture.

By 500 A.D., the chariot had gradually disappeared but, still beset with difficulties in raising good horses, cavalry had not replaced it to any great extent. There had been a return to the bow as the chief weapon of Indian warfare and many Hindu princes placed great reliance upon their powerful hill forts, which they used as bases for frequent raids on their neighbours. Although the barbarians who invaded their country were mainly horse archers, the Hindus never adopted this practice, probably because of the lack of good horses.

During the early years of the 11th century Mahmud of Ghazni, leading mobile forces formed largely of mounted archers, invariably defeated the far more numerous but static Hindu armies. It was a period when the numerous North Indian princes fought a series of wars against each other, using armies formed largely of ill-trained, badly disciplined and controlled infantry with the bow as their main weapon.

To sum up, Indian armies rarely showed any tactical sense, fighting without order and discipline. Few leaders of note emerged and the men they led, although courageous, were 'soft' in a military sense when compared with soldiers of other countries of the Ancient and Medieval world.

*Fighting Assessment Chart:* INDIANS

| | *Factor* | *Rating* | *Points* |
|---|---|---|---|
| 1. | Ability to assault in controlled order | Average | 1 |
| 2. | Bravery | Average | 1 |
| 3. | Control | Average | 1 |
| 4. | Defensive ability | Average | 1 |
| 5. | Desperation | Not applicable | – |
| 6. | Discipline | Average | 1 |
| 7. | Ferocity | Not applicable | – |
| 8. | Manoeuvrability | Average | 1 |
| 9. | Mobility | Average | 1 |
| 10. | Movement, speed of | Average | 1 |
| 11. | Order | Average | 1 |
| 12. | Tactical Sense | Average | 1 |
| 13. | Training and Drilling | None | 0 |
| 14. | United action, capability for | Average | 1 |
| 15. | Weapon Handling | Average | 1 |
| 16. | Demoralisation, likelihood of | Likely | 0 |
| 17. | Impetuous Action, likelihood of | Likely | 0 |
| 18. | Panic, likelihood of | Likely | 0 |
| 19. | Reluctance to take definite action | Likely | 0 |
| 20. | Unscheduled Manoeuvres, likelihood of | Likely | 0 |
| | | *Fighting Assessment* | 12 |

# THE CRUSADERS

During the 9th, 10th and 11th centuries various warlike Eastern peoples were thrusting westwards towards Constantinople, the centre of the Mediterranean world. The strongest of them, the Seljuk Turks, erupted into the decaying remains of the First Moslem Empire to establish themselves as 'Protectors of the Caliph of Bagdad' and, driving the Greeks out of Armenia, were soon in control of almost the whole of Asia Minor including Nicaea on the very threshold of Constantinople. The Byzantine Emperor Alexius Comnenus, conscious that he could never recover his lost dominions in Asia without Western aid, appealed to the Pope, stressing that the Seljuk Turks were now in possession of the Holy Sepulchre so that it would be a Holy War. In 1095, from Clermont in Auvergne, Pope Urban II sent out a call to the Christians of Europe to avenge the Holy Sepulchre. The priests of every town and village in Europe preached fire and brimstone to both high and low, inflaming them with the desire to fight for the Cross. An unmentioned inducement to crusade was the chance of plundering the reputedly rich lands of the East! In this way the First and most successful of the Crusades was born, uniting Christians more than ever before or since.

The men who went crusading were a cross section of 11th-century feudal Europe, possessing an assortment of languages and backgrounds that made it impossible for them to be commanded by any one man, thus they were split into four great divisions each under its own chosen commander. The first, formed of southern Frenchmen and Italians, was under Raymond, Count of Toulouse, a veteran warrior who had fought against the Saracens of Spain. The second army, composed of 70,000 northern Frenchmen and Germans, was led by Godfrey de Bouillon and his two brothers Eustace and Baldwin. Eustace was an unenthusiastic Crusader, Baldwin was openly seeking the estates he lacked in Europe whilst Godfrey was said to be valiant and perhaps the least self-seeking of all the leaders of the Crusaders. This army followed the Danube, through Hungary and Bulgaria. The third army, the main one, contained the majority of the northern French contingents and a few English Normans, led by Robert of Normandy, elder son of William the Conqueror; Hugh of Vermandois, brother of the King of France; Robert, Count of Flanders and Stephen, Count of Chartres, Blois and Troyes. The fourth army, which probably saw more action and gained more glory than the others, was led by one of the greatest warriors of the day, Prince Bohemond of Tarentum. Composed of Normans and Italians, the army was actually engaged in besieging Amalfi when it became infected by the urge to crusade so that the siege was raised and everyone marched off to Constantinople. En route to this assembly area, the Crusaders had to fight off attacks by the Nomads of Central Europe incensed by this invasion of pillaging soldiers accompanied by hordes of camp-followers who devastated their country as they moved towards Constantinople. This vast assembly of fighting men and camp-followers (said to number as many as 600,000) reached Constantinople in May 1097.

After a brief stay, the host marched out and besieged Nicaea for seven weeks before leaving the town behind them with the Byzantine flags fluttering from its towers. On July 1st they fought the first important battles of the Crusades when the column under Bohemond came out of the mountains on to the plain of Dorylaeum to find a mighty Saracen army awaiting them. At once the Crusaders deployed into line and charged the masses of Turkish horse, who were trotting across the plain to the accompaniment of rolling kettle drums and clashing cymbals. The Saracens wheeled away and scattered before the charging knights, then galloped round their flanks, pouring in flights of arrows as they rode. The plain was a patchwork of horsemen—charging, wheeling, rallying, pursuing.

For the next three hours, the Crusaders made headlong charges that rarely came to grips as the light Turkish cavalry melted away before them, to wheel into their flanks in darting attacks accompanied by a constant shower of arrows. Indeed, lacking an infantry screen, if the Crusaders halted then swarms of Saracen horse archers raced across their front goading the Western horsemen into yet another desperate and fruitless charge. Although their actual losses were not particularly heavy many of the Crusaders had lost their horses and were reduced to fighting on foot. Realising that he could not beat this Arab horde on his own, Bohemond sent for help to Godfrey and Raymond commanding the second column moving on a parallel road. They arrived when Bohemond's force was tired and demoralised by their inability to come to grips with the fast-moving Saracen horsemen amid the swirling dust-clouds. Set upon in flank and rear, the Turks found themselves caught between the two bodies of Crusaders, unable to manoeuvre or to withstand the heavy Western cavalry crashing through their ranks, overthrowing their horses and hewing their riders down with heavy swords and axes. The Saracens faded away in utter rout at some points where they suffered their heaviest casualties—claimed by the Crusaders to have been as many as 30,000 out of an army of 360,000 Turks, Persians, Saracens and Africans. Although these figures were undoubtedly exaggerated, it was a definite victory for the Crusaders. Nevertheless, with hindsight, it is possible to see in the victory at Dorylaeum the beginning of the pattern of warfare which was to lead to the decisive Saracen victory nearly a hundred years later at Hattin. Undoubtedly the Saracens were surprised to be defeated whilst both sides were amazed at each other's courage and fury. The Crusaders had taken heavy losses through fighting, disease, the heat, hunger and thirst but these very privations together with their victory served to weld them into a more cohesive force despite their different nationalities, tongues and leaders. They had painfully acquired a wholesale respect for the Turks and, whilst the Turks similarly respected the Crusaders, it is most likely that this first encounter convinced them that they held tactical advantages which would give them victory in the long run.

It has been said that a Crusading army resembled '. . . a vast and shambling shanty town' (Henry Treece). Tradesmen, prostitutes, priests, children and beggars in their thousands trailed after the army like a cloud of locusts. Within the army itself the great lords were often accompanied by their wives and families, who in turn were attended by maids, servants, cooks, pages, clerks and musicians. Just as each lord had his personal retinue so the knights and even the lower ranks were accompanied by their followers on a lesser scale. The warriors who made up this motley host were of many different types but the backbone of the army was the armoured horsemen, the high-born, the landowners or those who held it in fief from some liege lord to whom they owed service. Their preoccupation with hunting and fighting made them natural warriors, physically strong and well trained with the weapons of the day, riding trained war horses and wearing costly suits of protective mail. More than a match for any number of the ill-armed and unarmoured common folk, at the time of the First Crusade these knights were using arms and armour essentially the same as that used by the Normans at Hastings thirty years before.

His coat of mail (hauberk), with elbow-length sleeves and falling just below the knee and divided between the legs so as to make riding more comfortable, was made of leather or strong linen on to which were sewn numerous flat rings of iron. Under this was worn a gambeson, a long tunic of wool or linen with sleeves down to the wrists, often stuffed with wool and quilted so that it absorbed some of the force of blows which, though they might not pierce the chain mail, would otherwise have driven it into the flesh of the wearer. Thick stockings (chausses) usually made of linen and cross-gartered with leather thongs covered the knights' legs, some horsemen followed the fashion set by Duke William at Hastings and wore chausses of mail. On his head the Crusader knight wore a conical helmet with a projecting and protective nose-piece, often over a mail hood or sometimes a leather cap-piece to prevent the iron helmet chafing his head. Off-duty, he sometimes wore a cap of soft felt of any colour except yellow (worn only by the Jews at this time). His metal shield, reaching as high as his shoulder, was kite-shaped so that he could use it on horseback to cover all of his left side from neck to ankle, when he was using a lance. This lance was made of wood, often ash or pine, with an iron head and a metal shield (vamplate), covering the handgrip. His sword was long and straight —the extra long, two-handed swords were used when fighting on foot and were rarely carried on horseback. Instead of a sword or lance some knights carried a hammer, axe or mace and everyone wore a dagger on the right hip.

Dressed similarly to a knight the ordinary soldier or sergeant would wear a less elaborately made hauberk, and he would be forbidden a knightly sword. He was armed according to the resources of his leader but often equipped himself with some other weapon of choice picked up on the battlefield. He rode a rounsey or pony of no particular breed whilst the knights rode well bred destriers ('Great Horses')—a horse somewhere between a shire-horse and a modern hunter in size and power. The archers, pikemen and other foot soldiers were rarely given any important function, usually following their mounted masters and finishing off wounded or dismounted enemy whilst the knights careered valiantly onwards. Following the Saracen habit, the Crusaders soon learnt to cover their iron armour with a linen coat to ward off the sun's heat on the metal and sometimes wound a length of cloth around their helmets for the same reason.

Later came the complete covering of chain mail with the head and face protected by the pothelm although the protection of the hauberk was increased by the addition of chausses. With their more protective armour, their heavier weapons and more powerful horses, the Crusaders had an advantage over the Saracens. But this was neutralised by the tactical pattern of the conflict between East and West where the Crusaders made a heavy thrust only to meet a resilient and fluid defence—literally the feather-bolster parrying of a blow by a heavy club. Time and time again the heavily armed and armoured Crusaders found their charges melting into thin air as smaller, nimble-footed Arab horses were wheeled and pivoted to fade before the charge, to retire, rally, threaten, break-up, charge and wheel again in a confused pattern made even more bewildering by the dust-clouds that masked their movements. And all this to the nerve-racking whistle of the arrow-clouds put into the air by the bows of the Saracen horse archers.

It is possible that these archers threatened more than they actually accomplished because it was only at close range that the light arrows of the Saracen bows could inflict much damage. A Moslem chronicler records that the gambesons of the Crusaders cross-bowmen were stuck 'so full of arrows that they looked like hedge-hogs' while their wearers were unharmed. Throughout the course of the Crusade the Western knights were killed or captured far more often when they became dismounted after their horses had been killed by the archers rather than being harmed by the never-ending clouds of arrows. Some of the Western infantry were armed with the crossbow which, although much slower in its rate of fire than the light Saracen bows, was far more powerful.

Basically, the real difference between the Crusaders and the Saracens was in their military tactics. Although neither were disciplined enough to carry out more than the simplest tactics, both to a large extent retained their characteristic methods of attack mentioned earlier. At first, success for the Crusaders depended upon their initial charge in a tightly packed bunch, sweeping all before them by their weight and lance, swords or axes finishing off the job at close quarters. If it got home it was an invincible charge and if the cavalry were allowed time in which to re-form, by their archers or pikemen mopping-up in the area of the charge, then the Saracens could rarely stand before them. But the Saracens rarely hung obligingly around to take these charges so the Crusaders were forced to adapt themselves to a style of warfare that combined the shock of heavy cavalry with the fire-power of crossbowmen. (Later the Crusaders used mounted archers in the same way as the Saracens.) Combining steady infantry armed with spears and axes with this new shock/missile method, the Crusaders were able to prevail over the light cavalry tactics of the Saracens to frequently defeat far larger forces.

Both sides showed outstanding physical courage, the Western knights fought with immense fury but naturally as bull-terriers do and with a characteristic overbearing self-confidence which contrasted with the fatalistic streak of the Easterners so that frequently small forces of Crusaders were able to put to flight many times their number of Turkish opponents. In hot climates and under exciting conditions it is possible for hysteria to spread rapidly, a fact which led to some recorded instances of Crusaders becoming berserkers as at Antioch to which the Crusaders marched after their victory at Dorylaeum. It is recorded that, during a particularly bloody engagement when the Crusaders were being attacked strongly from the rear, Bohemond gave a great berserk roar and called on all to stake their lives on a last desperate effort. So ferociously did he lead the counter-attack that the densest body of Turks, surprised and terrified by the storm that struck them, were massacred during their rout. After capturing the city, the Crusaders were themselves besieged within it by Karboga of Mosul and his 200,000 Moslems. Conditions had become desperate within the city when a humble priest, Peter Bartholemew, alleged that Saint Andrew had told him in a vision that the actual Roman lance which had pierced Christ's side as he hung from the cross was buried in the church of Saint Peter at Antioch. On digging in the church, a corroded rusty lance head was found that was immediately accepted by the Crusaders as the Holy Lance and a token of inevitable victory. Bearing the relic, the Crusaders marched like sleep walkers out of the main gate of Antioch, to be attacked on all sides by the Turks who were seemingly unable to turn them back either by weight of numbers or power of their weapons. Unnerved by this hypnotic berserk attitude the Turks drew off just as the Crusaders cried out that they could see riders on white horses carrying banners and led by Saint George galloping down from the mountains. So vivid were the emotions aroused by this vision that the Turks were thrown into a panic and swinging round their horses they tried to gallop away only for the Crusaders to cling to them, slashing and clubbing until the Saracen army fled in demoralisation.

One of the most interesting features of the Crusades, perhaps unique in military history, were the religious military Orders who were sufficiently powerful to disregard the word of the king and yet fought so fiercely against the Saracens that any of them captured in battle were immediately executed. The first of them, the Hospitallers, began as a monastic Order permitted by an enlightened Egyptian Governor of Jerusalem to establish a hostel for poor pilgrims. When Jerusalem fell to the Crusaders the Hospitallers were so uniquely acquainted with both Christian and Moslem ways of life plus a fine working knowledge of Moslem military dispositions and tactics that they decided to transform themselves from a corps of humble helpers to the pilgrims and become a military force that could retain the hold on the Holy City and keep open its routes. Although the Order still contained brothers to tend the sick, they became more notable as an army of knights, bound together by vows of poverty and chastity with the avowed intention of destroying Islam wherever their white cross-adorned surcoats could plunge into battle.

In 1123 was formed another similar Order, the Knights Templar, with the active support of the King of Jerusalem who was shrewd enough to see that in the Templars he had a permanent standing army on whom he could call at any time, differentiating from the great majority of Crusaders whose main object was to acquire the maximum amount of plunder in the shortest possible time before returning home to their estates and castles in Europe. Forming themselves into three classes, the Templars recruited far and wide until they formed a force of ambitious knights with the joint aims of keeping open the pilgrim routes to Jerusalem whilst defending its king. The knights, men of noble birth, wore a red cross on a white tunic. The sergeants, who wore a red cross on a black tunic, were middle-class warriors who also acted as grooms and bailiffs to the Templars. Religious, medical and non-military clerics formed the third degree of the Order.

So quickly did the glamour of the fierce-fighting Templars spread that young noblemen came from all corners of Europe to enlist in this Order of arrogant white-coated horsemen. As time went on the Templars and Hospitallers, used in the role of a permanent standing army, were settled in fortresses along the borders of the territory held by the Christians. Dedicated and disciplined, the two Orders generally provided at least half the Christian striking force in the field, as at Hattin. By now more than the mere commanders of military forces, their Grand Masters were so high in status that they sat at the treaty table alongside kings and princes. They were the most powerful land owners and long before the end of the 12th century they were the most significant military factors in the Middle East through their ability to build and financially maintain great fortresses at strategic points and to be constantly reinforced by idealistic recruits from the best families of Europe. Indeed, they were superb 'policemen' of the routes leading to the Holy places. But inevitably the aura bestowed by their courage and tenacity became sadly tarnished by a jealousy of each other that, in the end, led both Orders to adopt military tactics of an almost suicidal nature in their efforts to better or spite each other. For instance, in 1153 during a Christian attack on Ascalon, the Templars, determined that they alone should have the credit for victory, actually drove back with the sword their fellow Christians whilst 40 knights of the Temple rushed through a breach in the city walls only to be hacked down to a man by the Moslem garrison and their corpses suspended from the walls of the city. Their reckless and furious manner of fighting made the Templars and Hospitallers men to be feared and, holding it as a point of honour that they should never give or ask for quarter, they were allowed no mercy when captured.

Although both Orders suffered grievous losses at the Horns of Hattin in 1177, they seemed to stand apart from the slow but permanent disintegration of Christian power in the Holy Land. They were still fighting at Gaza in 1244 when, after leaving Jerusalem to its fate, their large contingents were wiped out. At the same time in Acre, the Templars and Hospitallers each put 300 knights in the field under the command of their Grand Masters but were completely defeated with both Grand Masters killed and only 33 Templars and 26 Hospitallers surviving. After that what remained of the Templars and Hospitallers were left to hold on to their few remaining strongholds in Syria, where they feuded among themselves and died against overwhelming numbers of Moslems. In 1266, the Templar castle of Safed which dominated Galilee was besieged by Egyptians under Baibars and, after a month of siege, the Syrian infantrymen deserted the castle leaving only the Templars to hold the walls. Desperately holding off the besiegers with sword and axe during the long and burning hot July day, the Templars became exhausted as the fighting allowed them no time to rest or prepare meals. At last they reluctantly accepted Baibars' terms of permitting them to retire in peace to Acre but on opening the gates they were all beheaded. Three years before Baibars behaved in the same manner at the Hospitaller fortress of Arsuf when a third of the 270 knights who held the fortress had been killed, the rest surrendered only to be ruthlessly treated. In 1270, Sultan Baibars, vowing to finally destroy both Templars and Hospitallers, broke into the fortress Krak des Chevaliers, the greatest Hospitaller stronghold. Dispirited, the majority of Templar fortresses then surrendered without even a token struggle. After the loss of Krak des Chevaliers, the Order of the Hospitallers declined in importance, possessing only one castle at Marquab and only able to maintain 300 knights in comparison with the 10,000 of its great days. The Templar still held a number of castles and although diminished was still capable of financially supporting itself. Perhaps the final death agony of the Templars and Hospitallers and indeed of the Crusades took place in 1291 when both Orders stood side by side vainly attempting to hold off the overwhelming Arab assault on Acre, the last symbol of Christian rule in the East. When the walls fell the massacre began and, on the Sultan's orders, all remains of the Orders were wiped out. Finally, the Templars only held the little island fortress of Ruad off the coast from Tortosa where they stayed until 1303. Later both Hospitallers and Templars set up a base on the island of Cyprus. Outside the period of this chapter, the Hospitallers (by then better known as the Knights of Saint John of Jerusalem) lost the Island of Rhodes to the Turks in 1522 but in Malta in 1565 9,000 Hospitallers (of whom only a few hundred survived) successfully resisted a force of some 40,000 Turks.

The Teutonic Knights, an Order of German warrior-monks, bought themselves a castle at Montfort in the Holy Land which they re-christened Starkenberg. Later their headquarters were maintained in Acre, but lacking the spiritual fervour of the Templars and Hospitallers, the Teutonic Order returned to the Baltic where it was given castles and extensive territories in exchange for help in taming the Prussians and other tribes.

Another interesting military aspect of the Crusades was the numerous sieges that took place as the missile-flinging siege-engines—the mangonel, trebuchet, scorpion, Perrier, Arblast and Espringale—battered away at stout walls whilst Beffroi's immense wooden towers dragged up against the wall of the castle, unloaded their crews of warriors. On their way to Jerusalem in 1097, the Crusaders besieged Marra, a strong fortress south of Antioch, constructing a tall wooden siege-tower which was set against the city walls over which it hovered, enabling the Crusaders to fling down rocks on the defenders, to fling out hooks and grappling irons in efforts to drag down defenders by their barbs. Other attackers, clumsy in their heavy armour, laboriously scaled tall ladders set against the walls. When Jerusalem was reached, the initial attempt to storm the walls was beaten off so the Crusaders settled down to a siege. Using wood from forests as far as 30 miles away and great balks of timber which had been laboriously dragged from Constantinople, two towers, each of three stories, were built. Higher than the city walls, they were great lurching constructions pushed and towed on wooden wheels by unprotected foot soldiers. Fire arrows, rocks, balls of blazing pitch and a Greek fire (a mixture of flaming asphalt and crude petroleum) was rained down upon these men and on to the timbers of the tower in an effort to set it alight, often foiled by wet horse hides laid on the wooden surface. On the top-most platform of the tower were archers who swept the wall while from the floor below the storming parties poured across drawbridges allowed to rest on the battlemented walls of the city. Pushed forward by the men behind them, the front ranks of the storming party were often hurled down into the moat or on to the rocks below but the attackers were constantly reinforced by troops climbing up long ladders from the base of the tower. Many of these towers used at various points on the wall stretched the defenders out so that counter-attacks were impossible and a foothold was obtained. Once the city had fallen rape, looting and the most ruthless reprisals were the order of the day. Sometimes the situation was reversed when the Crusaders were besieged, as in 1270 when King Louis of France leading an army almost entirely consisting of paid mercenaries and not religious soldiers was trapped in Carthage by a Tunisian army. Crazed with thirst because the wells had been poisoned before Louis walked into the trap, the Crusaders suffered such Eastern refinements as hot sand blown into the city on the prevailing wind and the rotting bodies of dead men and horses pitched over the city walls by Moslem siege-engines.

In July 1099 Jerusalem fell to the Crusaders who also succeeded in carving out kingdoms for themselves in the form of a narrow strip of coastal territory defended by a long line of fortresses guarded by small permanent garrisons mainly of Knights of the Temple and Hospitallers. There were also some more ports, including Acre captured in 1104. The kingdom of Jerusalem held its own for 45 years after its first creation but by 1144 the Seljuk Turks had reorganised themselves and were providing a strong opposition. In 1146 a Second Crusade started out but it was infinitely less successful than the First. From 1169 onwards, Saladin began to unite the Moslem world and in the fierce fighting that followed, the Saracens recaptured the Holy City of Jerusalem in 1187. This provoked the Third Crusade, a very grand affair led jointly by the Emperor Frederick I ('Barbarossa'), King Philip Augustus of France and King Richard ('Coeur de Lion') of England. By this time Crusading was regarded with some scepticism by the common people who learned that only the nobles struggled back after a defeat and the common soldier was beheaded or left behind in slavery. When ransom money was required for the nobles, it was the poor people who were taxed. The Third Crusade was a failure, Jerusalem was not retaken and Saladin's power was not shaken. King Richard of England became an almost legendary hero on the battlefield but failed lamentably in his dealings with both allies and Saladin over the treaty table.

The great days of the Crusades had ended but, exalted by the new Pope Innocent III and seeking to gain from the decadence of the Byzantine Empire, the Venetians strongly supported a Fourth Crusade. It lasted from 1200 to 1204 and did not reach the Holy Land at all although mercantile Venice gained much of the coast and islands of the old Eastern Empire. There were another three or four so-called Crusades and men dreamed of recapturing the Holy Land until the middle of the 15th century when the Crusading concept slowly and quietly faded leaving Eastern Christendom firmly in the hands of the Moslems.

Although a failure, the Third Crusade is probably the best known and apparently the most glamorous of them all, even if only for the colourful tapestry provided by Saladin on the one hand and Richard of England on the other. The moral effect on their troops of such leaders was immense in a period when a successful commander meant the difference between life and death. The battles of this period were fierce, hand-to-hand conflicts lasting only a short time and often they were poised on a commander's personal courage, his tactical ability, his insight into the enemy's intentions and the manner in which he used ground. Personal courage and a reputation for great bravery were incalculable assets and Richard was not only a skilful commander but possessed the physique and courage, the arrogance and treachery that made him fit only for warfare. His handling of the Battle of Arsouf in 1191 indicates that Richard was capable of controlling his army in a manner unusual for the period. He was marching along the Palestine coast from Acre to Jaffa before turning inland to Jerusalem, so closely attended by hordes of Saracen horsemen that Richard had to sustain a battle array for 19 days. Maintaining admirable discipline during the march so that his men remained unbroken under the constant showers of Turkish arrows, Richard marched his army in a line parallel with the shore, his dismounted men-at-arms and armoured knights who had lost their horses forming, with crossbowmen immediately behind them, a solid base from which his mounted knights could make their sporadic charges when the Turks approached too close. Encouraging the Turks to close in during their attacks so that they could be charged by the slower-moving knights, Richard kept them under a stern discipline that prevented their attacks becoming a wild scramble, re-forming them for fresh blows and restraining them carrying out wild and senseless pursuits until stranded far from their own forces with blown mounts. Although the Saracens hovered for days around the tight, well organised column, continually firing their showers of arrows, Saladin was unable to tempt the Crusaders into breaking their formation. Finally he had to press home an attack and was soundly beaten by his combination of heavy cavalry, armoured infantry and missiles, all welded together in a cohesive and disciplined formation.

Describing the Battle of Arsouf in 1191 between Saladin and Richard Lion Heart, the Saracen chronicler Boha-ed-din says:

> The Crusading infantry marched between us and their cavalry, keeping as level and firm as a wall. Each foot soldier had a thick cassock of felt, and under it a mail shirt so strong that our arrows made no impression on them. They, meanwhile, shot at us with crossbows, which struck down horse and man among the Moslems. I noted among them men who had from one to ten shafts sticking in their backs, yet trudging along at their ordinary pace and did not fall out of their ranks.

In the Near East, the West learned a lot about the art of war, about sapping, mining, the use of siege-engines and Greek fire. From bitter experience they learned that an army formed of different nationalities or jealous feudal units would always be beaten by a unified army, well led and fighting with determination. Both from the Saracens and the Byzantines, they learned much about fortifications and military architecture as Henry Treece says in his book *The Crusades* (1962):

The Romans who first went into the Near East carried with them the concept of a castle as being little more than a single stone tower, square in shape and encircled by a single or double wall, closed in by a natural or artificial moat. Such as them as returned to Europe brought back a much elaborated architechural pattern, to which castles were set high upon a crag, the towers rounded to prevent mining at the angles, and in which wall after wall enclosed a series of turrets, each providing a line of fire upon the other and each therefore protecting the assault of the other. Furthermore, these new Eastern-style castles had projecting upper section or machicolation, which enabled the defenders to lean over their opponents and to counter-attack them with such refinements as molten tar and Greek fire, another product of the Crusaders. In Britain, such castles were first set up against the Welsh at the order of Edward I by an architect who had served all his apprenticeship in the Near East. He created citadels so reminiscent of Antioch as Harlech, Caernarvon and Conway.

## STYLE OF FIGHTING

Because of the great mobility of their Saracen opponents, the Crusaders discovered, at some cost, that it was essential to maintain a strong and solid infantry base from which to launch their cavalry attacks. The Franks, as the Moslems called the Crusaders, were supreme in this facet of warfare and no other military force in the world could stand up against equal numbers of their mailed knights and men-at-arms. After Saladin's decisive victory at Hattin, won by separating the Crusader cavalry from their infantry and then defeating each in detail, the Crusaders learned to devote more attention to closely co-ordinating the actions and movements of their infantry and cavalry. This led to some reasonably effective operations of their combined arms, such as Richard's great victory at Arsouf.

Richard and a few more enlightened Crusader leaders, realised the importance of relating fire and movement and whenever possible launched their overwhelming heavy cavalry charges after the Saracens had been shaken by volleys of crossbow-fire.

Learning from the Moslems themselves and from the Byzantines, the Crusaders began to make use of light cavalry for screening and reconnaissance purposes so that manoeuvres of envelopment and ambush became possible. Mounted fire-power in the form of horse archers was also developed and the crossbowmen favoured by Richard were particularly effective against Saracen horse archers.

Richard I, the lion hearted of England, was a heroic knight as well as a good tactical leader with a dominant personality that enabled him to control the undisciplined knights and barons from many lands who were under his command. He turned the Crusaders into a united army and, on the march to Ascalon in 1191, he demonstrated exceptional strategic and tactical ability together with staffwork and logistics far superior to that normally found in medieval armies. Richard, in his period, deserves to be rated as an 'Above-average' commander.

To sum up, the Crusaders were courageous fighters who handled their weapons well but suffered from being a mixture of many nations with consequent indiscipline and impetuosity. On top of this, they were feudal cavalry who expended their energy under a hot sun in wild charges, often chasing far more mobile Saracens in feigned flight.

*Fighting Assessment Chart:* CRUSADERS

| | *Factor* | *Rating* | *Points* |
|---|---|---|---|
| 1. | Ability to assault in controlled order | Average | 1 |
| 2. | Bravery | First Class | 3 |
| 3. | Control | Average | 1 |
| 4. | Defensive ability | Average | 1 |
| 5. | Desperation | Not applicable | – |
| 6. | Discipline | Average | 1 |
| 7. | Ferocity | Not applicable | – |
| 8. | Manoeuvrability | Reasonable | 2 |
| 9. | Mobility | Reasonable | 2 |
| 10. | Movement, speed of | Reasonable | 2 |
| 11. | Order | Average | 1 |
| 12. | Tactical Sense | Average | 1 |
| 13. | Training and Drilling | Average | 1 |
| 14. | United action, capability for | Reasonable | 2 |
| 15. | Weapon Handling | First Class | 3 |
| 16. | Demoralisation, likelihood of | Unlikely | 3 |
| 17. | Impetuous Action, likelihood of | Likely | 0 |
| 18. | Panic, likelihood of | Unlikely | 3 |
| 19. | Reluctance to take definite action | Unlikely | 3 |
| 20. | Unscheduled Manoeuvres, likelihood of | Likely | 0 |
| | | *Fighting Assessment* | 30 |

# THE SARACENS

The word 'Saracen' comes from the Arabic word Sharkeyn meaning 'Eastern People'—a Seljuk, Mameluke or Ottoman Turk, a Kurd or an Egyptian or a Persian. Under Mahomet (died in 632 A.D.) and the Moslem Caliphs who succeeded him, the Arab armies strove to force upon the Christian world the three alternatives of conversion to the Moslem faith—submission and the right to follow another faith on payment of a large tribute, or death. In 634 Caliph Omar, a notable soldier, went to war with Heraclius, the Eastern Roman Emperor, and on August 20th 636 a decisive battle was fought in a blinding sandstorm on the banks of the Yarmuk near the sea of Galilee. With the river at its back, the Byzantine army was forced into a retreat that became a bloody massacre. Jerusalem fell after a siege that lasted more than a year and the Moslem tide swept on to push the Byzantines out of Syria. Then taking Armenia the Saracens crushed the Persians and pressed on to conquer as far East as the Oxus and the Afghan hills. With the capture of Alexandria in 641, the Saracens had, within a few years, recaptured all that the early Persian conquerors had taken from them a thousand years before. By the year 700 Roman Africa was in the hands of the Arabs, 11 years later they occupied Spain and in the year 717 their Empire stretched from the Pyrenees to Central India. They were also hammering on the gates of Constantinople but it was denied them by the strength of its walls, the ingenuity of Byzantine engineers and the efficiency of the army and fleet. At last in 715 the Byzantines defeated the Saracen army under Suleiman and destroyed the Arab armies stranded when their ships were burned. The Byzantines under the Emperor Leo III (aided by the Bulgarians) had beaten a Saracen force split with dissension among their leaders and short of supplies.

The Byzantines considered the Saracens to be the most wily and dangerous enemy they ever faced and after Constantinople was saved in the 7th and again in the 8th centuries, the Byzantine army repulsed sporadic Saracen raids over Asia Minor. The Saracens' main advantage lay in numbers, speed of attack and fervour and on occasions they adopted arms and military practices of their enemies but still they could not stand up to the Byzantines. Leo the Wise wrote: 'Horse for horse and man for man the Byzantines were heavier and could ride the Orientals down when the final shock came.'

At the other end of their Empire a Moslem–Moorish horde of horsemen raided deep into France where they were confronted by Charles Martel ('The Hammer'). The armies were said to have faced each other for a week with Charles awaiting the arrival of reinforcements and the Saracen leader Abd-er-Rahman endeavouring to get his plunder away. The army of the Franks was composed largely of foot soldiers armed with shields, swords, axes, javelins and daggers—it was well trained but ill-disciplined and a large proportion of it consisted of ill-armed levies. The great strength of the Saracen army lay in cavalry equipped with lance and sword but they were encumbered by plunder-laden baggage mules which seriously reduced their mobility. The Franks took up defensive positions, reported by a chronicler as follows: 'The men of the North stood as motionless as a wall; they were like a belt of ice frozen together, and not to be dissolved, as they slew the Arab with the sword . . . vast of limb and iron of hand, they hewed on bravely in the thick of the fight. . . .'

It would seem that the Franks were drawn up in a phalanx or shield-wall which the repeated charges of the Saracen cavalry failed to shake. In a battle that seems to have resembled Hastings and went on until nightfall, the ranks of the Franks held firm and the Arab cavalry, their leader killed, retreated from Tours in the year 732, never again to hold any sway north of the Pyrenees.

It is not unreasonable to say that the decisive successes of the Saracens in the early stages of their military career can be attributed to fanatical zeal rather than any particularly outstanding tactics, training or weapons. Repeatedly there are examples in military history, even as late as the Dervishes in the Sudan in the latter part of the 19th century, of relatively savage warriors being made almost irresistible by their fervour and fanaticism. When the initial glow of religious zeal died away, then the Saracens began to be halted.

At the beginning of the 9th century two men emerged, one from the East and one from the West, whose achievements brought about a new balance of power in the Mediterranean area. They were the Caliph Haroun-al-Raschid and the Holy Roman Emperor Charlemagne. Haroun personified the last great Moslem triumph before the Seljuk Turks, 200 years later, erupted into the Middle East bringing with them new and belligerent concepts of Islam that provided the West with an excuse or reason for embarking upon the Crusades.

As they advanced into Asia Minor, the Crusaders were surprised to discover magnificent castles and fortified cities, luxurious tents, bamboo shafts of the spears, delicately engraved ironwork of the lances, the rondache or light shield and carrier pigeons used by the Saracens. It was not long before the more intelligent Crusaders began to respect the Seljuk Turk as a soldier who, like the Normans themselves, came from a long line of fighting stock. Speedily, they found out that the swift-riding Turk, on his light and mobile pony, was a subtle fighter, sliding to right and left, avoiding the shock of the heavily armoured charge. Skilful with the bow and arrow from a moving horse, the Saracen tactics were to deceive and split the attacking force by moving away so that they galloped in pursuit and then after out-distancing the heavily armoured Crusader on rocky or rising ground, turning and picking him off at leisure from outside the range of his lance. Although a handful at close quarters when using their finely tempered swords, the Saracens greatest success was achieved by the use of the bow with which they brought the Crusaders to the brink of disaster not only on their first encounter but many times in the future. The bow was a composite weapon 3 feet long when strung, with a pull of 118 lbs, firing an arrow just over 24 inches in length but effective range was about half this. It is said that a flight arrow from one of these bows could travel 480 yards and that the average archer could send the heavier war arrow at least 300 yards. Tests conducted in later years revealed that the arrow could penetrate a half-inch-thick plank at 100 yards with the head and shaft of the arrow passing for three or four inches through the wood.

The Saracen horsemen rode smaller and lighter mounts than the Westerners and wore a light and flexible chain mail that allowed considerable freedom of movement. This they covered with a linen coat because of the sun's heat on the metal and often wound a length of cloth round their helmets for the same reason. The chain mail was of excellent workmanship and beautifully adorned with gilding; the helmet had a nasal protection, carried a plume on the crest and was light and highly ornamented with gilding and damascene work in gold. They carried a round, small shield often intricately decorated and with a boss or point in the centre.

In 1169, Saladin became sole ruler of Egypt and brought a new element into the pattern of the Crusades. Not only was he a leader possessing great military skill and devotion to his religion but also he was a man of great courage, character and integrity. Under his able leadership the Moslem world, from Cairo to Bagdad, united to form a great army and inspired by the preaching of a Jehad or Holy War against all Christians, they had three years of constant success starting in 1174, only being halted by a temporary reverse in 1177 at Telljezer. In 1180, Saladin offered the Franks a truce but, with incredible stupidity, they persistently broke the terms of their agreement until, in 1187, Saladin marched against Tiberias on the Sea of Galilee with an army of 100,000 men, led by 12,000 Saracen knights. After an unsuccessful siege of the Kerak of Moab, the great fortress of Reynald, Saladin's greatest enemy, the Saracens rode along the Valley of Jordan to capture rich Tiberias.

To oppose the vast Moslem army, the Christians could muster no more than 20,000 infantry and a little over 1,000 knights and this was only done by sadly depleting the garrisons of the walled cities of Palestine. Led by King Guy, set on the throne by the Templars as their puppet, the Frankish army set out after Saladin. Ill-advised by the impetuous Templars and Reynald of Châtillon, at dawn on Friday July 3rd 1187 King Guy gave the order to leave their green and well watered encampment to take the waterless and dusty road to Tiberias. Carrying with them the True Cross (a fragment of wood encased in gold and studded with precious stones) the Christians trudged through the baked hills, as hot as an oven on that airless day under a pitiless sky from which the sun beat down with a merciless intensity. That evening they reached the waterless plateau above which the Horns of Hattin towered a hundred feet into the air. Saladin's Moslems were well established in the irrigated green valley below, jealously guarding it against those thirst-maddened Christians who attempted to reach the water. After causing their enemy great discomfort during the night by setting fire to the parched grass which covered the plateau, the Moslems moved into position and completely encircled the Christian camp.

Outnumbering the Christians by odds of 10 to 1, the Moslems began their attack at dawn to an accompaniment of horns and gongs. Maddened by thirst a squadron of cavalry under the Count of Tripoli charged down the hill at the Moslems who did not attempt to meet the charge but opened their ranks and let the horsemen through, then closed them again so that not a man returned. The Saracens made charge after charge up the hill and as the day wore on the Christian army became smaller and smaller until by evening only a few hundred knights, their lips cracked and tongues swollen, stood in a tight packed circle around the wrecked tent of King Guy. One last charge and the Moslem hordes rolled over the Christians, capturing King Guy and Reynald, with some other leaders. Saladin personally executed Reynald but kept the King prisoner for a year before releasing him, a broken man.

After the Horns of Hattin, Saladin was master of the Moslem world and three months later Jerusalem fell to his armies, its inhabitants being treated with a humane restraint completely opposite to that which would have been exhibited by the Crusaders had the circumstances been reversed.

At Hattin, the bulk of the Moslem army was composed of mounted archers, most of whom disdained to wear armour. There were some infantry called harbieh, equipped with reed lances tipped with iron, scimitars or captured Frankish swords and with small circular shields and daggers. There were also a number of Naffatin, foot soldiers who carried capatults and wicker baskets filled with naphtha balls. Placed in the catapults and set alight, these balls were hurled deep into the enemy ranks to burn fiercely on whatever surface they touched, being almost inextinguishable.

That momentous day on the Horns of Hattin put an end to Christian hopes and aspirations in the Holy Land. There were other Crusades and other notable leaders, such as Richard

Lion-Heart, but the decay of Christian influence in the area can be dated from 1187.

## STYLE OF FIGHTING

The Saracens were courageous, mobile and highly disciplined but, in contrast to the majority of their foes, they were 'light-weights' in the sense that their strength lay in mobility rather than shock-action. Primarily effective only in attacking, the Saracens rarely failed to take advantage of a gap in a battleline although they lacked the weight to deliver an effective blow by shock-action against a strong defensive force whilst they themselves lacked defensive standing power. They were masters of the 'hit-and-run' tactic and their high mobility together with the brilliant use of light cavalry for reconnaissance and screening enabled them to envelop and ambush on many occasions. They used mounted fire-power in the form of horse archers which was only really successfully countered when the Crusaders armed themselves with the crossbow. Then, realising that the formidable Crusaders under Richard would only be beaten by taking advantage of every possible factor, Saladin effectively combined fanatic Arab and Egyptian foot soldiers with Mameluke horse archers. When well co-ordinated under such leaders as Richard, the combined fire-power and shock effect of the Crusaders was more than a match for the 'light-weight' Saracens—light cavalry, lightly armoured and equipped with lance and sword. Before Saladin grafted new tactics and a new spirit into the Saracens, the Crusaders were able to take on armies of Saracens far more numerous than themselves.

The Saracens realised that the Crusaders had to maintain a solid infantry base from which to launch their overwhelming cavalry charges, so the Moslems invariably tried to separate the Crusader cavalry from the infantry and to defeat each in detail. A favourite tactic to this end was to entice the cumbersome Frankish cavalry to pursue them and then to turn and slaughter when they were spread out, afterwards going back to plague the static Crusader infantry by mounted horse archers.

*Fighting Assessment Chart:* SARACENS

| | *Factor* | *Rating* | *Points* |
|---|---|---|---|
| 1. | Ability to assault in controlled order | First Class | 3 |
| 2. | Bravery | First Class | 3 |
| 3. | Control | Reasonable | 2 |
| 4. | Defensive ability | None | 0 |
| 5. | Desperation | Not applicable | – |
| 6. | Discipline | First Class | 3 |
| 7. | Ferocity | Not applicable | – |
| 8. | Manoeuvrability | First Class | 3 |
| 9. | Mobility | First Class | 3 |
| 10. | Movement, speed of | First Class | 3 |
| 11. | Order | First Class | 3 |
| 12. | Tactical Sense | Reasonable | 2 |
| 13. | Training and Drilling | Average | 1 |
| 14. | United action, capability for | First Class | 3 |
| 15. | Weapon Handling | First Class | 3 |
| 16. | Demoralisation, likelihood of | Average | 1 |
| 17. | Impetuous Action, likelihood of | Rare | 2 |
| 18. | Panic, likelihood of | Average | 1 |
| 19. | Reluctance to take definite action | Average | 1 |
| 20. | Unscheduled Manoeuvres, likelihood of | Rare | 2 |
| | | *Fighting Assessment* | 39 |

# THE OTTOMAN TURKS

The Turks were a vast group of semi-nomadic Asiatic peoples, generally supposed to be the same people as the Huing-Nu, who for so long troubled the Chinese Empire and compelled the building of the Great Wall. Speaking the same language but divided by their wanderings into a host of tribes, they were driven by Mongol expansion towards the Eastern Mediterranean from Central Asia. Not a primitive people, but heirs to the two mature cultures of Islam and the Steppes, they were fiercely dynamic Moslems unaffected by any such qualities of chivalry or Christianity. It was these Turks that founded and sustained the one great and lasting Empire in the history of Islam.

Except on such occasions as when Alexander produced highly mobile cavalry used with tactical purpose, the native Asiatic technology of war has frequently been superior to that of the European's. Carrhae and Hattin were Christian defeats, Asiatic mobility triumphing over the heavily armoured European knight and man-at-arms of the Middle Ages. The victory of the Seljuk Turks over the Byzantines at Manzikert in 1071 opened the door for Turkish inroads into Asia Minor. Seeking a land in which to settle and spread the Moslem faith, many Turkish chieftains and their followers drifted westwards into attractive Anatolia, whose westerly position aided them in surviving the Mongol onslaughts whilst it served as a rallying point for others of the Moslem faith.

The Ottoman Empire, founded by Othman (Osman) in the early years of the 14th century, was the leading military and political power of South-East Asia from about 1300 to 1550, a period during which they drove westwards—against Europe on the two fronts of the Danube and the Mediterranean.

In 1299, Othman began open warfare against the Byzantines and in a war that smouldered for generations cleared them out of Asia Minor. Meeting little effective opposition as they swept over the countryside his cavalry quickly subdued all the interior although some cities held out for a few years like islands. Othman's son, Orkhan, organised the Ottoman Turkish State as he created a military system that made existing European armies appear as disorganised mobs. Simultaneously he fought the Byzantines and the Seljuks, capturing Nicaea and Nicomedia from the Greeks and Pergamus and the Province of Mysia from the Seljuks. In 1356 the Turks crossed the Hellespont into Europe, taking Adrianople, and masses of them flowed into the Balkans. Orkhan's successor Suleiman I died and was buried in Gallipoli, the first Ottoman Prince to be buried in European soil. It is said that his tomb 'invited the races of Asia to perform their pilgrimage to it with the sword of conquest'.

He was succeeded by Murad I who pushed into the Balkans threatening Constantinople. Alarmed Christendom talked of a crusade but did no more than talk, so eventually the Greek Emperor was compelled to recognise the Ottoman conquest. Murad pressed on into Europe, taking Monastir in 1380 and defeating the Serbs at the River Maritza in 1371 and Kossovo in 1389. He took Sofia in 1385, falling at last in the hour of victory at Kossovo in 1389 when he was stabbed in his tent by a Serbian patriot. His successor Bajazet I ('Thunderbolt') occupied Thessaly in 1391 and annihilated a Crusader army composed mostly of Hungarians at Nicopolis in 1396. But at Angora in 1402 his army was overrun by Tamerlane's hordes and Bajazet was taken prisoner. He died in captivity and his sons fought between themselves for his Empire.

Had Christendom been united this was the moment when the

Turks could have been driven out of Europe but England and France were still engaged in the 100 Years War, whilst internecine feuds were tearing Italy apart. So the Turks gained a lengthy breathing space and were allowed to re-form under such outstanding sultans as Meheat I and Murad II.

Beginning in 1444, a great alliance of Hungary, Poland, Serbia, Wallachia, Burgundy, Genoa, Venice, the Greek Emperor and the Pope, led by the Hungarian John Hunyadi, inflicted a series of crushing defeats on the Turks and compelled them to sue for peace. With many Turkish subject-peoples in revolt, the Christians within sight of the Turkish capital at Adrianople and a Frankish fleet in the Hellespont, Hunyadi, by making a truce, missed the second opportunity in a generation to destroy the Turks. He broke the pact in 1444 but was defeated at Varna and four years later, when his Wallachian allies deserted to the enemy, he was routed at the fateful valley of Kossovo. Little now stood between the Ottomans and the conquest of Constantinople, the main aim of Mehmed II who became Sultan in 1451. He did this in 1453 after encircling the city with an army of 100,000 men to besiege its 14 miles of walls garrisoned by 7,000 men. No Christian force attempted to raise the siege, indicating very clearly that the age of chivalry had passed. The chief military point of note was that the siege was a landmark in the history of artillery. The capture of Constantinople saw Turks at the height of their military greatness, inflicting resounding defeats on any European armies that opposed their advance in a manner that recalled the depressing experiences of the Crusades.

In their day, the Ottoman land forces were the most highly organised army in the world, possessing a professional nucleus of infantry and cavalry, giving allegiance to the Sultan as an individual in a manner comparable to the Feudal system. Savage and cruel to their enemies, they were highly trained and self-disciplined because of their religious zeal and strict sense of allegiance to the Sultan. They were first-class soldiers because they instantly obeyed their commanders with no thought of concern for their lives in battle. They could exist on barley and water for a long time without bread and wine. The bulk of the army was irregular militia, settled on land in return for military service when required. Mostly cavalry, they constituted the main mass of the army. Also there were hordes of irregular troops, infantry (Bashi-Bazouks) and cavalry (Akibi) who were unpaid and fought for plunder. The massive Turkish forces were mostly horsemen with the Spahis as the elite cavalry. Each Spahi was responsible for recruiting and training between 2 and 6 additional horsemen whom he brought to battle rather as a Western knight was accompanied by his 'lance'. The Spahis were the original Sultan's bodyguard, highly paid but lacking the Janissaries' peculiar recruitment and training. Their chief weapons were bow, lance and short sword and they wore no defensive armour. There were also gunners, smiths, armourers and marines hired for single campaigns and garrison troops, the Piyade.

The Janissaries were justifiably considered to be the most celebrated troops in the Turkish army, the Corps d'Elite. Taken from Christian parents mostly in the Balkan countries, at the age of 10 to 12 as part of a recognised tribute due to the Sultan, the Janissaries were trained in special communities and affiliated to a religious order of Dervishes, receiving in their monastery-barracks a celibate education which made them fanatical Moslems. They received the best possible physical education, being highly trained in the handling of their weapons so that they were masters of attack at the same time as they were experts in building fortifications, earthworks, ramparts and ditches and improvising defence. They were warriors who had neither family ties nor country and went into battle despising death and injury; they had a great deal in common with the Knights of the Hospital by whom they were regarded as worthy opponents. Belonging to a race whose only military tradition was that of cavalry, the Janissaries were professional infantry at a time when infantry were out of fashion in the West. Perhaps the last-ditch stand of the Byzantine legions had taught the Ottoman Turks to appreciate the potentialities of a good infantry force. At the height of their greatness in the first half of the 16th century, the Janissaries numbered between 12,000 and 15,000. Their basic tactical unit was the orta, which ranged in size at different periods from 100 to 3,000 men. The ranks of their many officers were taken from the titles of household departments—the Chief Soup Maker and the Chief of the Bloodhound Keepers. The commander of the Janissaries was the Agha.

Never heavily laden with protective armour, in the 15th and 16th centuries the Janissaries had a small round shield, a metal helmet shaped like a fez with a sharp point on top and possibly some light mail. Their original weapon was the short composite bow which they used with light arrows. Sabres and daggers were standard equipment and at various times they used slings, crossbows, javelins, lances, straight swords, pikes, axes, maces, scythes and whips. Each section of the Janissaries wore their own coloured uniform bearing the emblem of their corps—a wooden spoon—if the Christians were not frightened by the trappings of the Janissaries they were certainly impressed by them!

Each unit of the Ottoman forces wore its own distinctive uniform bearing its emblem. Armour was of the mail type and was topped by a turban-shaped spiked helmet with a sliding nose-piece. The Anatolian Turks wore both mail and breast-plates. The Ottoman equivalent of plate armour included fluted helmets, damascened in gold, brigandines (a kind of armour jacket, with exposed gold rivets) and heavy cloaks with luxuriant decorations. Since the scimitar was a weapon without defensive value, they also continued to use the small circular shield or target after it had gone out of fashion in Europe. In close, hand-to-hand work, the Turkish army was much less efficient than that of the Christians, who were able to risk their persons with greater boldness.

The Bashi-Bazouks were a motley gang of rascals—many of them were Christians such as Slavs, Hungarians and Germans—fighting for plunder. The Ottoman method of fighting employed these irregulars as cannon-fodder by sending them in first with a line of 'military police' advancing behind them to encourage any waverers with whips and maces—behind the police came the Janissaries waiting to cut down any fugitive who got that far. Often the Bashi-Bazouks did very well in the first shock of an assault but they were easily discouraged if not immediately successful and the hopes of plunder receded. Against well armed and disciplined defenders they were sometimes handicapped by their own numbers, getting in each other's way, and in scaling assaults most of the attackers could expect to be cut down before they reached the top of the ladder. But these irregulars worried the defenders and their corpses filled the ditches so that by the time the Janissaries were thrown in the defence had weakened and success frequently followed.

The Ottoman military machine was a monolithic structure suffering from marked defects in that it was particularly vulnerable to a war on two fronts; pressure in the West accompanied by any warlike attitude on the part of the Persians was always disturbing. Remarkably few, not only of the Ottoman generals but of the common soldiers, were Turks and many of them were not even Moslems. It was an inflexible and over-centralised organisation with the Sultan as an absolute despot whose supreme command cramped initiative in his generals.

STYLE OF FIGHTING

Composed almost entirely of light and heavy cavalry, backed by the relatively small force of Janissaries, Turkish armies of this period based their style of fighting on the well disciplined and controlled Janissaries, often operating from field entrenchments that provided a base of manoeuvre for the cavalry. These cavalry were either light irregular skirmishers armed with the bow or heavier (although lightly armoured) well disciplined horsemen—semi-feudal units composed equally of lancers and archers. In their hey-day, the Turkish cavalry armies and their combination of missile power and shock effect were a match for any other armies that came against them.

Like all Asiatic cavalry armies, they were extremely mobile and fluid, using hit-and-run tactics that could demoralise all but the finest infantry or cavalry of a similar type to themselves.

*Fighting Assessment Chart:* OTTOMAN TURKS

| | *Factor* | *Rating* | *Points* |
|---|---|---|---|
| 1. | Ability to assault in controlled order | Reasonable | 2 |
| 2. | Bravery | Reasonable | 2 |
| 3. | Control | First Class | 3 |
| 4. | Defensive ability | None | 0 |
| 5. | Desperation | Not applicable | – |
| 6. | Discipline | First Class | 3 |
| 7. | Ferocity | Not applicable | – |
| 8. | Manoeuvrability | First Class | 3 |
| 9. | Mobility | First Class | 3 |
| 10. | Movement, speed of | First Class | 3 |
| 11. | Order | Reasonable | 2 |
| 12. | Tactical Sense | Reasonable | 2 |
| 13. | Training and Drilling | Average | 1 |
| 14. | United action, capability for | Reasonable | 2 |
| 15. | Weapon Handling | Reasonable | 2 |
| 16. | Demoralisation, likelihood of | Rare | 2 |
| 17. | Impetuous Action, likelihood of | Average | 1 |
| 18. | Panic, likelihood of | Rare | 2 |
| 19. | Reluctance to take definite action | Average | 1 |
| 20. | Unscheduled Manoeuvres, likelihood of | Average | 1 |
| | | *Fighting Assessment* | 35 |

# THE SWISS

Towards the end of the 14th century two completely different types of infantry, the Swiss pikemen and English longbowmen, began to inflict defeat after defeat upon the proud chivalry of Europe. Slowly, reluctantly, it became evident that the days of the knights were numbered as the Cavalry Age died, and that the foot soldier was coming into his own again after a thousand years of neglect. But it remained necessary to change the concept that infantry could only stand on the defensive and show that they could act offensively even against cavalry.

Sandwiched into that short period between the end of the Cavalry Age and the revolutionising of known methods of warfare by the use of gunpowder, a citizen-militia of infantry soldiers from the Swiss cantons devised a system of weapon tactics, organisation and training that gave them military dominance over most European armies for two centuries. Secure in their mountain valleys, for many years the Swiss defied those who would be their feudal overlords; as early as 1291 they formed, for self-defence, a league of cantons against the House of Hapsburg. Eventually in 1315, their open defiance aroused Duke Leopold of Austria, who considered himself their liege-lord, to lead 15,000 knights and infantry against them, who were crushingly defeated by 1,500 infantry at Morgarten. From then, the Swiss did not suffer any serious reverses until Marignano in 1515.

Like the Romans before them, the Swiss learned in the school of hard experience. Commanded by sound if unimaginative veteran commanders and knowing only one relentless way of fighting, they carried the day against more brilliant but less redoubtable foes. By doggedly applying discipline, steadiness and rigorous training, principles sadly lacking in 14th century European warfare, the Swiss achieved a cohesion and manoeuvrability that was irresistible. Remorseless drilling enabled them to perform complicated manoeuvres and to be capable of marching, wheeling, changing direction and retiring without losing their alignment. They introduced marching in step to military music or the sound of the drum to relieve the monotony of drill and improve march-discipline.

Intensely patriotic, brave and dauntless and without the slightest chivalrous feelings that might permit quarter being given to a defeated foe, the Swiss fought with an appalling ferocity. In 1513, at Novara, the defeated German mercenaries were slaughtered out of hand by the Swiss, dying with the knowledge that to surrender to these ferocious mountain men only prolonged life for a few more hours.

One undoubted reason for their success that became a morale-destroying legend to their enemies was an unshakeable belief in their own invincibility—they had a proverb, 'God is on the side of the Confederacy', and the simple medieval soldiers of both sides could not, with impunity, contest the will of God. On occasions, Swiss morale was so arrogant and presumptuous that they attempted the impossible or embarked upon enterprises which could only result in great loss without much material gain.

On breaking camp, the Swiss took up battle formation and marched brigaded in fighting order to the scene of action ready to enter battle as soon as the enemy were encountered. He, on the other hand, needed time to bully and jostle an often mixed force of militant knights and reluctant peasants into a battle array. So, even before the battle began, the appearance of the already formed Swiss formations on the field had a marked moral and physical effect. The Swiss tactical system was so simple, like that of the

Spartans and the Romans, that it hardly ever varied and forced their enemy to fight where they and not he chose. Formed up for battle, they presented an impenetrable bristling hedge with the pikes of the front ranks sloped forward at an angle, with the lower end of the weapon under the instep. So long was the pike that the men in the second, third and even fourth ranks could present them over the shoulders of those in front of them so as to form a hedge of bristling steel. When they advanced, the pikemen brought their weapons to the charge and the whole mass of men advanced slowly forward to win their objective by 'push of pike'. It was said that a Swiss infantry unit could advance in formation at a fast pace with

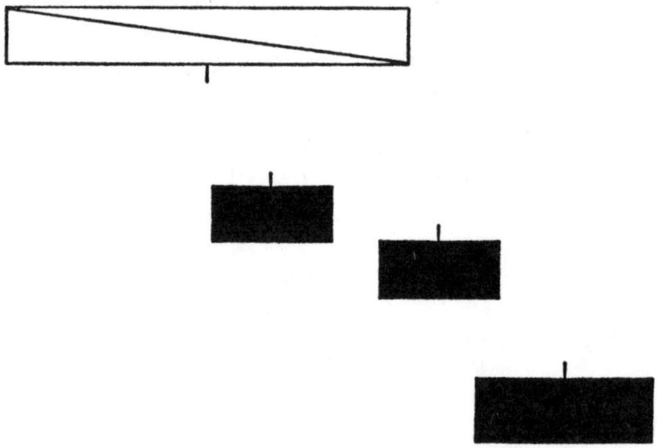

The attacking formation of the Swiss pikemen

their pikes held horizontally in both hands above their heads for more than a mile and then strike with almost the shock of cavalry.

Generally, the Swiss advanced in echelon of three divisions. Moving in step with their long pikes extended, fifes squealing, drums beating and flags waving proudly overhead, they would go forward without hesitation to attack even heavily armoured and mounted knights. The leading division (the Advance-guard) made for a pre-determined point in the enemy line, while the central division (the Main-battle) marched parallel with it, slightly to its right or left rear. The third division (Rear-guard), pikes held upright amidst a panoply of banners, marched further back and often halted when the battle was joined, so as to observe the first shock of impact before deciding on their best course of action. The three divisions were extremely fluid—sometimes the second division would lead, with both wings lying back (refused);

on other occasions the two wings would go in level with each other while the main division was refused.

The echelon formation prevented the enemy from wheeling inwards and catching the leading division in flank because, in attempting this move, the enemy would expose his own flank to the second Swiss echelon, which could move across to support the first. An added advantage lay in the open space in the rear of the attacking columns to which they might retire if repulsed, without danger of throwing troops behind them into confusion—a common occurrence in many medieval battles. At times the normal three-division formation was varied and on at least one occasion the Swiss moved in one large hollow square. When surrounded by cavalry, or by overwhelming numbers, they formed large squares with crossbowmen and halberdiers at the corners. Such a 'hedgehog' of pikes was formed at St Jacob on the Birs in 1444, when a force of less than 1,000 Swiss deliberately attacked an army of nearly 15,000 Armagnac mercenaries invading the Swiss Confederacy. They broke the enemy's centre and were soon surrounded by great numbers of horsemen but their square of pikes, although repeatedly charged by mounted men-at-arms and under constant attack by crossbowmen and light troops, maintained its position until nightfall amid heaps of their own dead and ringed about by some 2,000 Armagnac corpses.

Sir Charles Oman wrote of them:*

> The rapidity of the Swiss advance had in it something portentous; the great wall of pikes and halberds came rolling over the brow of some neighbouring hill; a moment later it was pursuing its evil way towards the front . . . the solid masses gliding forward in perfect order and in deep silence . . . and then—almost before the opponent had time to realise his position—it was upon him . . . the war-cry burst out in one simultaneous roar and the column, with its four rows of spear points projecting in front . . . dashed itself against the solid front . . . with the impetus of file upon file surging up from the rear.

Far in advance of its time, the Swiss tactical unity was an excellent example of deficiencies being turned into assets. Their sole aim of hurling a remorselessly advancing solid mass of 18-foot pikes at the enemy was the best possible way to compensate for no cavalry and no artillery-train whilst taking all possible advantage of the fact that, unlike other armies of the day, they were unimpeded by a mob of ill-armed and undisciplined feudal infantry. By attacking first and so forcing the fight, moving with

* *A History of the Art of War in the Middle Ages* by Sir Charles Oman (1959).

frightening speed and cohesion, they were able to take the fullest advantage of their own superior mobility and discipline, to give the maximum momentum of their attack whilst reducing to a minimum the length of time during which they were under enemy fire. Their invincible 'hedgehogs' revived the big fighting formations which had been dead for a thousand years. As the rest of Europe imitated them, the fate of battles began to depend upon infantry instead of horsemen. Soon there was not more than one man in ten on horseback instead of two out of three.

Because they could not afford to buy it, the Swiss wore little or no armour, quickly realising that the greater mobility so afforded them made it possible to out-march, out-manoeuvre and act characteristically upon the offensive against heavily armoured opponents. Against the slow-moving, sprawling medieval forces they encountered, their compact manoeuvrability allowed them to bring overwhelming force to bear at a given point. In his book *Art of War* Machiavelli praises the Swiss rapidity of movement, saying: 'No troops were ever more expeditious on the march, or in forming themselves for battle, because they were not overloaded with armour.'

The great majority of the Swiss wore only a cloth cap or hat and a leather jerkin. Those who wore a steel cap and a breastplate were stationed in the front rank. The leaders wore full armour and were often mounted until the action commenced when they dismounted and led their men on foot.

In its day, the pike, together with the English longbow, was a most frightening weapon. The Swiss were not the first to use the long weapon, even as early as the days of Philip of Macedon its potentialities were known but they were not fully exploited until the later Middle Ages. The Scots used pikes with considerable success against the English but the first people to really exploit the weapon were the men of Brabant, in Flanders. As early as the 13th century these Brabancons were hiring themselves out as mercenary bands in France and Italy, often being employed as a rallying point for cavalry or to bear the brunt of attack while the heavy horsemen were forming after an unsuccessful charge. Lacking the training, discipline, fervour and confidence of the Swiss pikeman of later years, the Brabancon similarly fought *en masse* but could not manoeuvre rapidly. On the move and in close order, the Brabancons kept their weapons raised in the air with those of the rear ranks passing over the heads and shoulders of the men in front of them. This meant that it was not possible to incline suddenly or turn to right or left without knocking over the front ranks or causing them to fall into disorder. As used by the Brabancons, the pike, being longer than the horseman's lance, was a deadly weapon and almost impossible to break in defence but weak in advance so that cavalry were required to work in conjunction with the pikemen, pursuing the enemy once the pikes had checked and broken them.

The first serious cavalry reverse of the Middle Ages was at Courtrai in 1302 where a Flemish infantry army of pikemen destroyed a French army of knights, when the horsemen's charge bogged down in a stream so that they were helpless against the long and heavy pikes.

The Swiss also used their national weapon, the eight-foot halberd, a heavy spear with a sharp-pointed head and a hatchet-like blade at the front, with a strong hook or spike at the back to drag men from their saddles. There were several types of halberd—bills, guisarmes, voulges, fouchards . . . all murderous weapons at close quarters. Some of them had blades almost 30 inches long and wielded by brawny Swiss arms these terrible weapons could shear through armour, flesh and bone like a knife through butter. The halberdiers were usually stationed at the centre of the column, guarding the banner; when the column was brought to a standstill, by a sudden foe, the halberdiers issued out of the side and rear ranks and slashed at the enemy flanks. Sometimes they were held in reserve for the final slashing attack. It was a Swiss halberd that struck down Leopold of Austria in 1386 and in 1477 laid open the head of Charles of Burgundy from crown to chin.

Some Swiss armed with two-handed swords or morning-stars (a heavy club studded with long spikes) were stationed behind the pikes to move forward between the files to the front rank if the onward surge of the phalanx was checked. The Swiss also used light troops armed with missile weapons—crossbows in the early days, but by the time of Granson and Morat (1476) the use of firearms was common. Usually, these missile men were deployed as skirmishers in front of the advancing pikes, falling back between the files to the rear of the formation before the opposing lines met. The proportion of light troops to the rest of the armies varied—at Morat they numbered 10,000 out of 35,000, although this seems to have been exceptional.

The Swiss never had a single commander-in-chief, conducting their wars through a council composed of veteran soldiers, captains of the forces from each canton, who discussed questions of policy and strategy and took decisions by majority vote. This system of command, frequently beset with jealousy and rivalry between the cantons, prevented the Swiss from strategically following up their victories so that they were never able to establish themselves as a first-class political power. Joint leadership by veteran commanders was most successful against the usual feudal armies they encountered—a veteran soldier with an eye for ground could be relied upon to lead his division of pikes at the weakest point in the enemy line without requiring any great knowledge of tactics or grand strategy. This destroyed the need for anything more than competent leadership because, with everyone thirsting to get at the enemy, it only remained to work out the quickest and most effective way of getting to grips. In time, although the Swiss had perfected their few old and practised manoeuvres, developing tactical situations occurring during battle could not be met by varying their plans because of that lack of flexibility and incapacity for improvisation inevitably resulting from their system.

The proud and arrogant Swiss refused to adapt themselves to the advance of military science and armament as the use of cannon and the musket made it impossible for unaided pikemen to succeed on the battlefield. The massed phalanx was at a disadvantage against good generalship, backed by powerful artillery and small arms, but the Swiss persisted in tactics which had been so successful against the inept blundering of feudal princes. At Marignano in 1515, the charge of the Swiss mercenaries was broken by the French cannon and their day was over.

Before this the commanders of every army in Europe had racked their brains for ways of defeating the Swiss formations. Perhaps the most successful were the Spanish who to give greater offensive capabilities mixed sword-and-buckler-armed infantry with their pikemen. After the pikes had clashed together, the Spanish swordsmen slipped between or under the low weapons of friend and foe alike in much the same way as the ancient Romans fought with their short thrusting sword many hundred years before. As his opponent closed with him in this manner the bearer of the long pike became helpless. The Spanish infantry were a formidable match for the Swiss pikemen during the 15th century. The Emperor Maximilian of Germany believed that the only way to beat the Swiss was to play them at their own game, so he raised regular corps of pikemen and halberdiers, called landsknechte. They were trained to act in exactly the same manner as the Swiss and some of the bloodiest conflicts ever known were fought out, with quarter neither being asked nor given, when they came upon each other. Then would two bristling lines of pikes cross and the leading ranks thrust upon each other's weapons by the pressure from behind, going down almost to a man at the first impact. When the long weapons had become hopelessly jammed together by the press of men from behind, the interlocked divisions were assaulted by the halberdiers of both sides who ran out trying to encircle each other and slashing inwards at the seething mass with their fearsome weapons. The side that was finally forced to retire was badly mauled in the process—the landsknechte are said to have lost over half their men in a combat with the Swiss at Novara in 1513. Showing the same mercenary tendencies as the Swiss, the Germans held their pike low down and pointed it upwards whilst the Swiss grasped it nearer the middle thrusting it slightly downwards.

Their tactics gave the Swiss a lengthy succession of brilliant victories—battles in which they won their independence from the Duke of Austria and maintained it against the Duke of Burgundy—Morgarten in 1315; Laupen in 1339; Sempach in 1386; Näfels in 1388; Granson in 1476; Morat and Nancy in 1477. At Morgarten an Austrian army, straggling along in the usual feudal disorder, was ambushed in a narrow, slippery alpine pass bordered on one side by a steep slope and with an icy lake on the other. After sending a torrent of boulders and logs down the slope into the packed Austrian ranks, 1,500 Swiss (outnumbered 10–1) charged down into the disorganised column wielding halberds, clubs and morning-stars. In a rout that turned into a butchery, many of the Austrians were forced off the path and into the lake. A contemporary writer said: '. . . for the mountain folk slew them like sheep in the shambles; no one gave any quarter, but they cut down all, without distinction until there were none left to kill'.

Terrain advantages and bad Austrian leadership gave victory to the Swiss at Morgarten but Laupen, 24 years later, was fought on an open slope with the outnumbered Swiss fighting in their three-column formation for the first time. Charging downhill at the enemy as they began to ascend, the Bernese on the right were at once successful but the Forest Cantons on the left had their assault halted by cavalry and were forced into their 'hedgehog' defensive formation, until relieved by the timely arrival of another column. At Sempach in 1386, 6,000 Austrians, dismounted by their commander who thought they would use their heavier equipment to better advantage on foot, were almost destroyed to a man by 1,600 Swiss.

The Swiss reputation for courage and obstinacy reached its peak during the 15th century in the war against Charles the Rash, Duke of Burgundy. At Granson in 1476, the Swiss vanguard came on to the battlefield well in advance of its two fellow divisions and, with characteristic arrogance, took on the entire 30–40,000 strong Burgundian army. After beating off a cavalry attack, the vanguard then not only held but began to press back the picked lances of the Duke's personal guard, considered the best in Europe. The arrival of the remaining two Swiss divisions caused the Burgundian infantry to break and flee in panic, abandoning their camp with a fortune in spoils to the Swiss. A few months later, the Duke of Burgundy came forward with another large army to meet the Swiss at Morat. Picking the moment to attack a few thousand Burgundians holding an entrenched palisade after the main army had retired to its camp, the Swiss divisions drove them downhill to disorder the rest of the army coming up in support. The battle soon developed into a rout and a pursuit made notable by the ferocious butchery of the fleeing Burgundians by the Swiss.

In this battle the Swiss had a small cavalry force made up of about 100 Bernese aristocrats whilst it is said that there were some mercenary English archers among the forces of the Duke of Burgundy but there is no record of them achieving any success against the Swiss—perhaps the rain in which the battle was fought affected their bow strings. The defeat of the Dukedom of Burgundy, reckoned a power in Europe, raised the prestige of Swiss fighting men to great heights so that their services were sought as mercenaries by all the Princes of Europe.

At Novara in 1513 three columns of pikemen in echelon, after a clever feint to mask their real intention, had made havoc of the French cavalry. The Swiss pressed on their victory as far as Dijon before they were bought off for a very large sum. The Florentine historian Guicciardini wrote: 'Never did the Swiss nation make a finer . . . resolve, a few against many, without horsemen or field pieces to attack an army so admirably provided with both. Many have ventured, considering the courage, to reckon this deed of ours more highly than most of the memorable actions recorded of the Romans.'

Novara was the last major victory of massed pikemen.

Towards the end of their period of fighting fame, the Swiss became the leading mercenaries of Europe, exporting men-at-arms in exchange for gold. They were professionals capable of taking the field at very short notice so as to be upon the enemy before he was aware of their presence. At the same time they had many infuriating peculiarities such as leaning on their pikes and refusing to fight if they found that the army opposing them contained men of neighbouring Swiss districts or leaving the field, even though the battle was about to begin, if their wages were not paid up to date. In the Italian wars at the end of the 15th century involving the French, the Spanish and the many Italian States, the Swiss fought indiscriminately for money on both sides. Perhaps the decline of the Swiss, generally attributed to the increasing use of artillery and hand guns, was accelerated by an insiduous lowering of character and discipline encouraged by the brutalising effect of their cold-blooded ferocity on foreign soil.

At Marignano in 1515 the French artillery of Francis I tore holes in the massed ranks of Swiss pikemen, allowing cavalry charges to penetrate their formations. Guns and Spanish infantry played major roles in ending the legend of Swiss invincibility. At Bicocca in 1522, artillery and an improved arquebus played havoc again with the Swiss so that in many ways the Italian Wars, like the Italian Renaissance, marked the end of the Middle Ages and the beginning of a new era.

## STYLE OF FIGHTING

The successes of Flemish and Swiss pikemen gave a substantial tactical superiority to defence over offence in European warfare. Through superb discipline and training, the Swiss pikemen could take successful infantry offensive tactics against cavalry without any particular dependence upon missile weapons. It was a sort of reincarnation of the Macedonian phalanx. Like the Macedonians, the Swiss knew that if their pikemen could maintain their massed formation without gaps and without faltering in the face of the cavalry charge and if they could press home their own attack with speed and shock, their massed pikes were more terrifying to the cavalryman than he was to the foot soldier.

In wargaming, they must be given a power that reflects the universal recognition that they were incomparably the finest troops in the world. In addition, their manoeuvrability, cohesion and shock-power must be given due weight and, perhaps more than anything else, their ferocity must be considered. Through sheer cold-bloodedness, the Swiss aroused in their opponents an almost umatched sense of terror (be sparing with 'Desperation' factors!).

Usually fighting in three columns of phalanxes echeloned to the left or right rear, they marched into action in that formation. If only one route was available, the first group marched directly into the fight with the others peeling off to right or left and rushing up on to the flank of the first column.

Their formations included a few crossbowmen as skirmishers

and a number of halberdiers in the interior files of the column whose job was to cut down individual horsemen in a mêlée, when pikemen had come into contact with a body of horse. On some occasions they used great numbers of skirmishers.

Brilliantly manoeuvrable, highly disciplined and trained, always under excellent control, the Swiss had a complete moral as well as tactical ascendency and, in their day, no other infantry in Europe could stand up to them.

*Austrians and Burgundians*

The usual foes of the Swiss, the soldiers of these two countries invariably lost when they came into contact with the ferocious pikemen from the mountain cantons. The Austrians and Burgundians fought in a feudal manner, with high-born knights and cavalry acting impetuously and without control whilst ignoring their infantry who they consistently failed to support. Ill-disciplined and badly trained, they showed courage but were invariably half beaten by the Swiss reputation before the battle commenced.

*The Spanish*

By the close of the 15th century Spain had the finest and best balanced military force in the world, mainly because they had moved faster than other nations in the exploitation of the use of gunpowder. At the same time they had developed techniques with pike and sword that made them the first army to achieve any success against the Swiss pikemen—that is until the increased use of the scorned firearms and artillery brought the Swiss to the ground.

In the second half of the 15th century, through constant campaigning against the Moors and by modelling their infantry on a contingent of Swiss mercenaries, the Spanish began to emerge as one of the finest armies in Europe.

In this period the Spanish were well-trained, excellently disciplined and controlled, courageous and apparently undeterred by the formidable reputation of their Swiss opponents. To include them in this book is to hover around on the fringes because the emergence of Spanish military supremacy really owes more to their diligence in applying themselves to learning the art of warfare with artillery and firearms.

*Fighting Assessment Chart:* SWISS

| | *Factor* | *Rating* | *Points* |
|---|---|---|---|
| 1. | Ability to assault in controlled order | First Class | 3 |
| 2. | Bravery | First Class | 3 |
| 3. | Control | First Class | 3 |
| 4. | Defensive ability | First Class | 3 |
| 5. | Desperation | Not applicable | – |
| 6. | Discipline | First Class | 3 |
| 7. | Ferocity | First Class | 3 |
| 8. | Manoeuvrability | First Class | 3 |
| 9. | Mobility | First Class | 3 |
| 10. | Movement, speed of | First Class | 3 |
| 11. | Order | First Class | 3 |
| 12. | Tactical Sense | Reasonable | 2 |
| 13. | Training and Drilling | First Class | 3 |
| 14. | United action, capability for | First Class | 3 |
| 15. | Weapon Handling | First Class | 3 |
| 16. | Demoralisation, likelihood of | Unlikely | 3 |
| 17. | Impetuous Action, likelihood of | Rare | 2 |
| 18. | Panic, likelihood of | Unlikely | 3 |
| 19. | Reluctance to take definite action | Unlikely | 3 |
| 20. | Unscheduled Manoeuvres, likelihood of | Rare | 2 |
| | | *Fighting Assessment* | 54 |

# THE HUNDRED YEARS WAR

At some time in their history nearly every race on earth had used the bow and arrow, but nowhere did they reach the pitch and skill and perfection as in England during the 14th and 15th centuries. In that period the English bowman dominated the wars of Europe as no comparable force has ever done since. The mere sight of them was enough to strike the fear of God into an enemy who, if he did not retreat or keep his distance, was almost certainly slaughtered. Somehow the enemy never learned this. French, Irish and Scots—each of them lost thousands of their youth and nobility to the laconic English bowman.

It was a time when England was a young nation, feeling her feet and still a little unsteady. The triple victories of Crécy, Poitiers and Agincourt stoked up the fires of national consciousness to forge a pride that has never left these shores. It is the stirring story of the Hundred Years War, of Edward III, of the Black Prince, and Henry V, but, more than that, it is the saga of the English archer, because without him, and the tactics built around him, none of the victories in France during the medieval period would have been possible.

Edward I, like his grandson and great-grandson, was an able soldier, capable of devising new expedients in war, also he possessed considerable strategical ability. Learning from his Welsh wars, this monarch introduced a more scientific use of archery that started the longbow's lengthy supremacy in war. His methods had been foreshadowed more than half a century earlier when in 1106, Henry I dismounted his knights and won at Tenchbriar (Tinchebrai) against Robert of Normandy and again in 1125 at Breuville in the same manner. At Beaumont he added a company of archers who brought the Norman cavalry crashing to the ground. These tactics were probably copied (like the order of battle) from a Byzantine model and taught the English the second of two most useful lessons (Henry had already discovered that dismounted knights could hold their own against the impetuous French knights) that a cavalry attack could be weakened, almost to annihilation, by volleys of arrows. Such knowledge, at a time when cavalry held absolute supremacy in war, was of unfathomable value and laid the foundations of England's military power.

Directed by brilliant, brave and far-seeing captains, the English army did not lose a major battle between Morlaix in 1342 and Patay in 1429. Along with the Swiss pikemen, the best professional fighting man of his day, the English yeoman and his longbow were the most significant single factor that changed all the old traditions and concepts of medieval fighting and warfare. It reduced war to two simple elements, one or both of which have to be employed to defeat an enemy—he must be overthrown either by shock, or by missile fire, or by both in combination. The shock method means that success if achieved when one side bests another, often through superior numbers, in a hand-to-hand struggle. This method is materially affected by the superiority of arms or the greater strength and skill with which they are wielded. The missile method means that the day is won by one side keeping up such a constant and deadly rain of missiles that the enemy are destroyed or driven back before they can come to close quarters—in this manner a smaller force can defeat a larger one. Both methods are capable of numerous variations and techniques and combinations of various arms and tactics.

In their simplest and most elementary forms the English archer and the Swiss pikeman represented these two basic methods of military efficiency. The former relied on his ability to beat the enemy by highly skilled shooting of great accuracy; the latter by being able to present a solid column faced with a formidable hedge of spear points able to drive before them superior numbers of the enemy who were unable to withstand the crashing impact and steady pressure of the pikemen. The common factor was that both methods overthrew the heavy mail-clad horsemen who had for so long been masters of the battlefield. Because of the marked superiority of the two methods they were copied by those who had suffered from them; but neither was easy to accomplish and no one ever succeeded in becoming better than the originators.

Archery was simply a primitive form of artillery, playing the same part then as now by softening up the enemy to allow the infantry to get to grips under the most advantageous conditions. There are numerous instances of archers being used in most intelligent fashion to cover the movement of other troops—a 'combined operation' at the landing of English men-at-arms at Cadzand, in 1346, enabled them to get comfortably ashore whilst the defenders were pinned down by hails of arrows from massed archers. In the autumn of 1342 Robert of Artois was besieging Vannes, then second City of Britanny; after spending the first few days in making preparations for the assault D'Artois delivered it early one morning. The archers put down what would now be called a 'standing barrage' on to the battlements—so fierce and accurate that, according to Froissart, the battlements were soon cleared and not an enemy dared show his head. Covered by this fire, the men-at-arms advanced to the assault.

The early 14th century saw the evolution of a coherent military system which used in a single tactical scheme the distinctive power of archery, the defensive solidarity of dismounted men-at-arms and, when necessary, the offensive power of mounted troops. Edward I realised the virtues of archery in attack to break up a defensive infantry formation and its power in defence when based on a formation of dismounted knights and men-at-arms. A beginning was made in the Welsh wars: at Orewin Bridge in 1281 when the men of Prince Llewellyn prepared to stand their ground in a defensive position as the English advanced against them, archers interposed with cavalry. The arrows inflicted sufficient loss on the Welsh troops to cause them to loosen their cohesion and fall into comparative disorder so that the cavalry were able to ride them down. At Maes Maydog in 1295, near Conway, the Earl of Warwick used the same tactics. A contemporary report says:

The Welsh on the earl's approach, set themselves fronting his force with exceeding long spears, which, being suddenly turned towards the earl and his company with their ends placed in the earth and their points upwards, broke the force of the English cavalry. But the earl well provided against them, by placing archers between his men-at-arms, so that by these missile weapons those who held the lances were put to rout.

In later battles with the other hereditary enemy, the Scots, the effectiveness of combined archery and cavalry action against immobile infantry formations was shown.

The pattern of the tactics that were to bring success against Scots and French was seen at Falkirk, in 1289, when Edward I led an army consisting of English, Welsh and Irish infantry, with a powerful body of mailed, mounted and well disciplined cavalry. They probably numbered 10,000 foot and about 1,500 horse. To oppose them William Wallace, the Scots leader, collected a force largely composed of peasantry organised as spearmen and armed with pikes some eleven foot in length, plus a group of archers from the Ettrick Forest and a force of about 500 cavalry. They were considerably less in number than the English army, but they had great confidence in their leader, who had positioned them in a very strong site to await battle.

That the Scots were formidable opponents is borne out by the very discerning opinion of the archer in Conan Doyle's book, *The White Company*:

'I have heard that the Scots are good men of war,' said Hordle John.

'For axemen and for spearmen I have not seen their match,' the archer answered. 'On the other hand, they are the sorriest archers that I have ever seen, and cannot so much as aim with the arbalest, to say nought of the longbow. Again, they are mostly poor folk, even the nobles among them, so that there are few who can buy as good a brigandine of chain mail as that which I am wearing, and it is ill for them to stand up against our own knights, who carry the price of five Scotch farms upon their chests and shoulders. Man for man, with equal weapons, they are as worthy and valiant men as could be found in the whole of Christendom.'

Their position was fronted by Darnrig Moss, a marsh through which no cavalry could pass; the flanks had been skilfully protected by wooden palisades driven deep into the earth and roped together. The Scots spearmen were formed into four great masses (schiltrons) of circular form and ready to face outwards in any direction. The spearmen levelled their long pikes from the hip to repel cavalry; the immediate front ranks kneeling on the right knee, against which the butt of the spear was placed so that a bristling wall of sharp spear points presented itself in whatever direction the cavalry might choose to attack. Between each of the schiltrons was placed a band of the Ettrick archers, whilst the cavalry remained close at hand in reserve.

The English army attended Mass, celebrated by the Bishop of Durham clad in full armour with a sword by his side and a shield slung at his back. The array that surrounded the militant churchman was impressive—the banners bore the arms of Edward . . . gules, three lions passant regardant; and those of St Edward the Confessor—a cross fleury between five martlets or. The tunics worn over the mail shirts were elaborately painted and blazoned; those curious ornaments called ailettes were worn on the knights' shoulders. The barrel-shaped helmets were surmounted by their crests; skull-caps, spherical and conical, were worn by the infantry; the lances had little emblazoned banners hanging from their heads. The Scots' banners showed the Scottish lion rampant. and the silver cross of St Andrew; Wallace himself wore a helmet surmounted with a dragon crest.

At the King's orders, the archers concentrated their rain of arrows on particularly stubborn points of resistance and clothyard shafts poured unceasingly into the unarmoured ranks of the Scots bringing them crashing to the ground. Before long the 'wood of spears' began to waver as man after man dropped to the ground with an arrow projecting from his unprotected body. Deserted by their cavalry and with their own archers destroyed, the Scots infantry levelled their pikes over a breastwork of their own dead and dying and made desperate attempts to hold their ground. Here and there a man crept away from the formations; soon the trickle became a stream bringing disorder.

Edward threw in his cavalry to thunder across the torn ground in a sudden charge, lumbering through the scattered ranks, laying about them with lance, sword and axe. Wielding his great two-handed sword, Wallace slugged his way from the field accompanied by a handful of faithful followers. The monumental stubbornness of the schiltrons at Falkirk in 1289 proved a match for cavalry charges far more violent than anything seen in the Welsh Wars. Fire-power, in the form of archers, supplemented by cavalry could beat such tactics, but this formation was to be the basic method of infantry fighting up to the early 19th century at Waterloo, outlasting the heavy-armed cavalry and long surviving the introduction of firearms.

With the notable exception of Bannockburn, for the two centuries that followed the Battle of Falkirk, its characteristics were almost monotonously repeated whenever Scots and English met in battle. Halidon Hill, Neville's Cross, Homildon and Flodden were all variations on the same theme. If the Scottish spearmen stood firm, they were decimated by archery until the English men-at-arms came into the assault. If the Scots attacked, they were beaten by dismounted men-at-arms, flanked by archers. In these early years the essential military conditions of success in the Hundred Years War, both in tactics and in organisation, had been devised.

In spite of having a population three or four times larger than that of England, France was never able to put into the field an army capable of standing up to the English forces until the latter stages of the Hundred Years War. Until then she did not succeed, or even try to change her policy of raising troops by the feudal system, so that the French armies typified all the bad military qualities of the period. Invariably, they were the usual feudal host composed of jealous lords and their retainers, scorned infantry and the usual local levies of ill-trained infantry who were strengthened by foreign mercenaries such as the Genoese crossbowmen, and occasionally aided by bodies of courageous Scots carrying on, in different fields, their perpetual struggle with the English. At the very beginning of the war Crécy proved the stock example of French disdain for any form of co-operation between aristocratic cavalry and all other despised arms, it was a lesson that France took nearly a century to learn.

Because the feudal organisation of society made every man of gentle blood a fighting-man, but not necessarily a soldier, a feudal army presented an unbelievable collection of unsoldierlike qualities. Although arrogance, stupidity and great courage coloured the activities of these armies, their inability to replace skill and experience made tactics and strategy impossible. The knight had no conception that discipline and tactical skill were as important as courage; it was always possible that at some inopportune and critical moment a battle might be precipitated or a carefully laid plan ruined by the incredibly foolish bravery of some petty knight with lust for only personal glory. Social status rather than professional experience led to command, so that the noble with the largest following was always superior to the skilled veteran with only a few lances to lead.

They recognised no superior but the King, and, unless he were a leader of uncommon skill, he was often powerless to control them, so that the radical vice of insubordination continued unchecked. Their very formation encouraged this in many ways; confined to a single pattern, they were formed into three great masses or battles and then launched at the enemy; there could really be no other way because the troops were neither disciplined nor accustomed to act together so that combined movements of small bodies were impossible.

Keeping a reserve in hand was a refinement practised by very few commanders, partly because it would have been very difficult to persuade a feudal chief to stay out of the front line of battle so that he incurred the risk of missing some of the hard fighting. Regarded as a model of military efficiency if he could sit his charger steadily and skilfully handle a sword and lance, nothing could restrain him when the enemy came in sight. His shield would be shifted into position, the lance dropped into rest, the spur plunged into the charger and the mail-clad line would ponderously roll forward. Thundering on as they gained speed, they had little regard for anything that might lie before them; as often as not the formation dashed themselves against a stone wall or tumbled into a ditch; painfully floundered in a bog or surged futilely around a wall or palisade. If the enemy were similar to themselves the two forces would meet with a fearful shock, men and horses tumbling in all directions, and then a chaotic mêlée would follow, sometimes lasting for hours. This meant that most engagements were nothing more than a huge, sprawling scuffle and scramble of men and horses over a patch of bare land or a hillside. Sometimes, as if by general agreement, both parties would laboriously wheel to the rear, halt for a while as their horses regained breath, and then rush at each other again until one side was worsted and fled from the field. The most elementary military tactics, such as pre-selecting a battle position, or using a reserve to take the enemy in rear or flank, were considered examples of exceptional military skill. The commendation of the age boiled down to striking individual feats of arms rather than any efforts at real leadership.

Great battles did not take place very often, simply because opposing armies often completely lost each other through neglecting to keep in touch by vedettes or outposts and patrols. It was usually the existence of some topographical objective, such as a road, ford or bridge, which precipitated a conflict; with maps non-existent and geographical knowledge both scanty and inaccurate, it was easy for armies to stray away and lose sight of each other. A recognised manner in which this last contingency was prevented lay in the opposing generals solemnly sending and accepting challenges to meet in battle at a given place and a definite date.

There was little for the infantry to do, no important part for them to play; they accompanied the armies for no better purpose than to perform the menial camp duties and assist in the numerous sieges of the period. Now and then, as a sort of overture, they were used to demonstrate ineffectually at the opening of a battle, but if they presumed to prolong these demonstrations, their lords, affronted by such presumption, would end the skirmishing by riding into and over their luckless followers. The half-armed peasants and burghers who had unwillingly joined the levy because it was the duty of every able-bodied man to do so were incapable of combining to withstand a cavalry charge; lacking adequate weapons and without discipline, they were ridden down and crushed.

The only infantry who commanded any respect were those bodies who were armed with more or less uniform equipment and weapons; the chief cause of the military unworthiness of infantry generally can be said to have been due primarily to the miscellaneous nature of their armament. The Scottish lowlanders, with their long spears, and the Saracen auxiliaries, plying their crossbows, stood out as troops capable of putting up good performances on foot and without all the benefits of high birth. The few infantry successes which occurred towards the end of the feudal period were exceptional and served to foreshadow the new era of co-ordinated dismounted warfare.

When a feudal host came up against a force or a commander capable of exercising even the most simple and rudimentary tactics on the field of battle they invariably took a hiding. With each commander making his own speed into the attack, followed by his supporters, the feudal force arrived at the scene of battle in small scattered groups. This meant that the battle was made up of a number of detached and unco-ordinated cavalry combats and a systematic enemy could defeat each of these groups in detail so that the sum total of the small routs added up to a great defeat.

By introducing auxiliaries like the English archer against a military caste too hidebound and blind to alter its losing methods throughout almost the whole of a hundred-year period, the English commanders could hardly fail to bring to earth the flower of French chivalry.

The English infantry were Welsh spearmen and archers. Although the longbow originated in Wales it soon crossed the border and first Cheshire men and then archers of other counties were armed with it; all the archers in Edward III's army were Englishmen. They were strong, muscular men; tall, sinewy, brown, clear-eyed and hard-visaged—middle-sized or tall men of big robust build, with arching chests and extraordinary breadth of shoulder. Their profession was proclaimed by the yew or hazel stave slung over their shoulder, plain and serviceable with the older men but gaudily painted and carved at either end when belonging to younger archers. Steel caps, mail brigandines, white surcoats with the red Lion of St George, and sword or battleaxe swinging from their belts completed the equipment. In some cases the murderous maule or five-foot mallet was hung across the bow-stave, being fastened to their leathern shoulder-belt by a hook in the centre of the handle. When they went to war, spare bow-staves were taken, plus three spare cords allowed for each bow and a great store of arrow-heads.

English archers carried into the field a sheaf of twenty-four arrows, buckled within their girdles. A portion of them, about six or eight of them, were longer, lighter and winged with narrower feathers than the rest.

An archer was often called upon to shoot straight and fast; but often he had to deal with an enemy hiding behind a wall or an arbalestier with his mantlet (a wooden shield) raised—the only way in which such protected men could be hit was to fire in such a manner that the shaft fell upon them straight from the clouds. Even as early as Richard I's siege of Messina, the archers drove the Sicilians from the walls in this manner—'for no man could look out of doors but he would have an arrow in his eye before he could shut it'.

It is evident from the fact that they wore so little defensive armour that the archers were designed to be light infantry, swift and mobile, skilful and deadly with their weapons. The English archer was reputed to be able to draw and discharge his bow twelve times in a single minute, at a range of 250 yards, and if he once missed his man in those twelve shots he was but lightly esteemed.

For a weary and sick army of less than 6,000 men to defeat over 25,000 French at Agincourt must indicate that the archers could notch with a shaft every crevice and joint of a man-at-arms' harness, from the clasp of his bascinet to the hinge of his greave.

English archers tried to avoid fighting with the sun in their front, considering the dazzling splendour of a summer's day to be very unfavourable to shooting. At Crécy, when the sudden gleam of sunshine after the rain burst forth behind the English, its beams, besides dazzling the eyes of the enemy, flashed upon their polished shields and corselets with a lustre so brilliant that the archers discharged their first flight of arrows with more than usual certainty of aim.

In addition to being incomparable with his missile weapon, the English archer would frequently discard his bow and fight on foot with sword, axe or maul. When a knight was seated on a horse it was almost impossible to get any power into a swing with a sword, so that he had to stand up to deliver his blow. Standing in the stirrups, he left exposed the one unprotected part in his whole armoured body—his seat. This was the target of the nimble archers and they seldom missed with their keen swords as they dodged on light feet in and out of the horse and foot mêlée. Well could Sir John Fortescue say: 'The might of the realme of England standyth upon archers.'

The English army had lifted itself from the dragging chains of the feudal system to become a paid, professional short-service army in which the mounted noble and the yeoman archer served overseas under contract to the King. It was a highly trained and disciplined mercenary army; a soldier drawing regular pay for his services is more amenable to discipline than the man dependent on looting and plunder. Edward's army was the most powerful and highly trained force of its day.

In the matter of arms and armament the soldiers of both countries were not dissimilar. Both sides had men-at-arms (knights were men-at-arms but men-at-arms were not necessarily knights) armed similarly with sword, lance, dagger and sometimes battle-mace, helm, shield and spurs completing the equipage. The knights had three armed attendants, who might be pages, to clean and polish his armour, help him in and out of it, hold his horse and assist him to mount; they also groomed the horse; then he had two mounted archers and one swordsman; the whole constituting a 'Lance'. He also had three or four horses, including two heavy chargers (destriers). From the end of the 13th century the horses themselves wore defensive armour. A chanfron protected its head whilst the neck was covered with a crinet with mail attachment and the front of the horse's body was protected by the peytral, its sides by the flanchards and its rear by the crupper. A strong horse had no difficulty in carrying this defensive covering which in the later stages of its development only weighed just over seventy pounds, including saddle and mail.

Men-at-arms were covered in armour from top to toe (cap à

pie). The increase in plate-armour reduced the mobility of the men-at-arms as it reduced the effectiveness of the arrow. There is plenty of evidence in the chronicles of the French Wars that if men-at-arms, covered completely in plate, advanced against English bowmen without their too-vulnerable horses then they would stand at least some chance of coming to handstrokes. When a body of fully armoured men-at-arms plodded with bent heads into the storm of arrows, however powerfully the shafts struck the hard, smooth, curved surfaces of the armour, they would glance off unless they found lodgement where plate overlapped plate. The effectiveness of plate-armour gradually made the shield unnecessary—it was ineffective against cannon-balls, anyway. Armour improved slowly from about the middle of the 13th century when mail was worn, with a flat-topped barrel helm; then from about 1280 it was reinforced with plate and the helm was 'sugar-loaf'; from 1300 there was further plate reinforcement, and a visored helm (the great bascinet), and in the 15th century complete plate-armour was common—this was undoubtedly the finest period of armour. French archers were armed with the crossbow, more powerful than the longbow but four arrows could be fired in the time it took to discharge one bolt. Usually used by Genoese mercenaries, it was more inaccurate and had a shorter range.

Up to now the longbow had been employed principally in defensive warfare against an enemy numerically inferior in cavalry. But when Edward III led his invading force into France the conditions of war were entirely changed in that they were up against a country invariably superior in the numbers of their horsemen. The yeoman with his longbow was soon to find that the charging squadron presented an even better mark for his shaft than the stationary mass of infantry formed by the Scots schiltron. At the beginning of the Hundred Years War, in the early 1340s, the Continental world had not yet learned that it was almost hopeless for cavalry to try to force, in a frontal attack, a position defended by men-at-arms supported on their flank by archers. The Battle of Halidon Hill formed the prototype for Morlaix, the first pitched land battle of the Hundred Years War and all the other great battles of the war—except the last. At Morlaix, the Earl of Northampton was victorious, against odds of 4–1, using Halidon Hill-style tactics that were to be followed by the English throughout the war—the men-at-arms fighting dismounted, the trench in front forming an obstacle (a marsh at Halidon); the defensive position on a ridge, the skilful use of the archer's fire-power in co-operation with supporting heavy troops.

Almost for the entire period of the Hundred Years War, when the French attacked on horse, they packed their men-at-arms into a close and solid mass with ranks and files closed up as tightly as possible to maintain a compact array. Even if the English archer was not always able completely to prevent the French attack striking home, he was able to decimate its ranks so that it was weak and disordered when it reached the English position. A man-at-arms was not a headlong galloping cavalier, his attack could not be very rapid unless it was made in disorder; it was shock-action, but shock of a ponderous column moving at a moderate rate. Moreover, if the arrows did not penetrate the armour their effects were such as though they did, for the presence of archers in the field eventually compelled the French to advance on foot. Though plate-armour is not much heavier than mail and is most flexibly jointed, it is not meant for marching in. The necessity of having to trudge a mile or more, often uphill or over ploughed land or through long grass and scrub (as at Mauron in 1352 and at Poitiers in 1356), and to fight at the end of it, was almost as devastating to the French men-at-arms as having his horse shot from under him. More often than not, he died in either case. It is most marked that in all the English victories during the Hundred Years War it was always the French who attacked and trudged up the hills in their armour. The English quietly stood about, waiting in their strong, carefully chosen defensive positions, perfectly fresh for combat when the exhausted Frenchmen came to grips with them.

In the summer of 1346 Edward marched his army from the Cherbourg Peninsula almost to Paris, crossed the Seine at Poissy and marched towards Amiens almost to make contact with his Flemish allies. On 26th August 1346, the English army formed up on a windmill-crowned ridge at Crécy and without stirring from that position, completely defeated the French army of 40,000, killing a third of them. The English army, between 12,000 and 13,000 strong, was positioned in three distinct divisions. That of the Black Prince, consisting of 800 dismounted men-at-arms flanked on either side by a total of 2,000 archers and about 1,000 Welsh spearmen, was placed well down the slope within 300 yards of the valley bottom. On the Prince's left, and somewhat drawn back so that they were slightly higher up the slope, lay the rearguard under the experienced Earl of Northampton; it was smaller than the first division, consisting of about 500 men-at-arms and 1,200 archers formed up in the same manner. The right of this division rested on the Prince's left and its left flank was protected by Wadicourt. The third division, that of the King, consisted of 700 men-at-arms, 2,000 archers and perhaps 1,000 Welsh spearmen, and were formed on the plateau in front of the wood of Crécy-Grange, behind the battle of the Prince of Wales. The baggage was parked in a wagon-leaguer backing on to the wood; its interior being occupied by the horses and garrisoned by the pages. Edward intended to fight with his men-at-arms dismounted, like Northampton at Morlaix a few years previously.

The men-at-arms were deployed into line by the marshals and

English long bowmen of the Hundred Years War period formed up in a 'harrow' or wedge-shaped formation

then a solid wedge of archers (Herce) formed up on the flanks of each of the divisions, with archers inching forward diagonally, pivoting on the flank of their own men-at-arms; where the two contiguous lines of archers met, an apex was formed. In this way a bastion-like formation was created in the intervals between the divisions; obvious advantages being that the front of the men-at-arms and the flanks of the army could be enfiladed by arrow fire. The archers dug small holes in front of their position and planted a plentiful supply of arrows in the ground; their usual supply of twenty-four or forty-eight arrows being supplemented from the wagons. When they were exhausted, the archer had three choices:

Small groups forming their archers in front of the men-at-arms in groups each of five men

(1) He could await the arrival of a fresh supply from the wagons.

(2) He could dash forward during a lull and pick up arrows fired at the enemy that had missed their mark and were lying on the ground. This was done at Poitiers.

(3) He could abandon his bow and join in the mêlée with his sword, as he did at Agincourt. Being unencumbered with armour, the archers were more nimble and most effective in hand-to-hand fighting. They were probably hefty, muscular men, as only a strong man could effectively wield a longbow.

Crécy proved that the archer, when supported by dismounted

men-at-arms, could beat off the most determined cavalry charges. For generations the French persisted in the delusion that the defeat was due to the stability of dismounted English men-at-arms. In part this was true, because the newly successful English tactical scheme depended upon men-at-arms fighting dismounted and in mutual support of the archers. Undoubtedly, the qualities of the two armies had a great bearing on the result of the battle; on the one hand the English were well trained, well led, well disciplined and well armed. The French, on the other hand, were a hastily collected force from different countries, not particularly well trained and no unit knowing much about its neighbours. As a result it lacked cohesion, neither trusting nor respecting each other, so that, as an army, it was bound to disintegrate when exposed to ordered blows.

The Battle of Crécy marks a step in the progress of the military art, in the age-long contest between mounted and dismounted men, between missile and personal weapons and in the emergence of a third arm: artillery. Edward had laboriously transported in wagons some crude iron tubes firing iron and stone balls—the first appearance of cannon in open warfare, it appears only to have had a minimal effect.

The Battle of Poitiers (1356) was the second of the great trinity of memorable victories, of which Crécy was the first and Agincourt the third. Although its details are somewhat controversial, it is fair to claim that it is easily the most interesting and instructive of them all from a military point of view.

Formed up on a ridge, bounded by a hedge, thick with scrub and undergrowth, the English army was about 6,000 strong, composed of 3,000 men-at-arms, 2,000 archers and 1,000 sergeants; the force was positioned along or close to the hedge. The archers, dismounted, were for the most part drawn up in the familiar Crécy formation—on the flanks of their respective divisions, in wedges slightly in advance of the men-at-arms. The Prince's division were in reserve in rear, and he also kept back a small body of mounted men. The French army was about 20,000 strong; formed in four bodies by the order of the French King, the men-at-arms were dismounted and had cut their lances down to five feet for close fighting.

It was a transitional period so far as armour was concerned; mail was gradually being replaced by plate-armour. The Black Prince wore plate except for his mail gorget, but the ordinary knight wore a bigger proportion of mail, esquires had even less plate than the knights. Over the breast-plate or hauberk was worn a spectacularly emblazoned and loose-fitting surcoat, bearing the arms of the knight who wore it; this gave a recognised rallying point for his followers during any crisis or emergency in the battle. As the strength of armour increased, so the need for the shield diminished—its size decreased until it was eventually discarded. The archers, both foot and mounted, wore a steel cap and breast-plate or a padded hauberk. Spearmen were similarly attired, except that they rarely ever wore a breast-plate.

During the battle, seeing the approach of the French cavalry, the English archers on the left of the position intelligently moved still further leftwards, into the marsh. In the waterlogged ground they were comparatively safe from the heavily clad horsemen, and, undisturbed, were able to keep up a galling shower of arrows into the French flanks. The English arrows were ricocheting off the French breast-plates as the riders advanced directly towards the archers; the new position enabled them to fire obliquely at the unprotected hindquarters of the horses rather than at the armoured riders.

First beating off cavalry attacks, the English men-at-arms fought a hard hand-to-hand battle with superior numbers of dismounted French men-at-arms and routed them. Then came yet another advance by a huge French column—and the archers had few arrows remaining. The Black Prince mounted his men and charged down-hill into the French force, which was eventually routed when struck by a small force of cavalry sent out to come in on the French flank.

Poitiers was a throw-back to the previous century in that it was won by men-at-arms with a successful late action of mounted against dismounted troops. Although it began in the Crécy tradition, the archers did not really play their usual prominent part.

In 1402, a Scots force had fortified themselves on Homildon Hill and were forming themselves into schiltrons when suddenly they were attacked by some 500 English archers, sent forward on reconnaissance. Bodies of Scots spearmen attempted to charge down the hill, but all were thrown back by showers of arrows. Under the galling hail, the national temperament of the fiery Scots rose to boiling point; they began to move from their defensive position and to surge down the hill. When the surge became a flood, the English archers commenced to retire in sections, discharging volleys in succession, causing the baffled spearmen to lose men at every step as they impulsively pressed forward. A body of Scots horse lumbered to the front but, in the same way, were unable to get into close quarters. It was maddening and ridiculous that this small body of men should be causing such heavy casualties without losing a single man of their own.

The Scots attempted to re-form their normally close-packed ranks into a defensive formation when they saw the massed ranks of the English awaiting them, but were still being ruthlessly assailed by the hissing clothyard shafts of the archers. These Scottish pikemen in close-knit groups and unprotected, if they had any armour at all it was only the very lightest, fell in dozens. Their leaders, who still used the old English chain mail, found it no defence against the deadly arrows; horses wounded and ungovernable, their breasts and flanks bristling with bloody arrows, galloped madly to and fro, trampling both dead and wounded into a gory pulp.

Percy and March, the English commanders, now sent in the English mounted men; with a shout, the archers dropped their bows and took to their swords and axes, rushing in and mingling with the horsemen, to play their final part in the mêlée. There was not a great deal for the men-at-arms to do. They drew their swords and laid their lances in rest, but the Scots had had enough; broken by the archers, they turned and fled.

The Battle of Agincourt was fought in 1415 between 6,000 half-starved and weary English (5,000 of them were archers) and the flower of the chivalry of France, about 25,000 strong. Monstrelet, in his chronicles, describes the English archers as being for the most part without armour, and in jackets with their hose loose, without hats or caps and often barefooted; their hatchets or swords hung at their girdle. St Remy says that they were not bareheaded, and that many of them wore caps of cuir bouilli, or boiled leather, and others of wicker-work crossed over with bars of iron.

Filling one side of a rectangle some 940 yards wide and bordered by woods, the English faced the French force across an area of muddy plough. The small English army was drawn up in the usual three 'battles' of dismounted men-at-arms with archers on the wings of each battle and a further two bodies of archers, one on each wing of the army. The men-at-arms were about four deep and the archers about seven to the yard, being formed into wedges or bastion-like formations projecting in front of the line of men-at-arms.

The French army, mainly composed of men-at-arms, was formed in three lines, all dismounted except the rear one; there were two bodies of cavalry each 600 strong, stationed on each wing. The lines were about five or six deep, although the French front of about 1,200 yards was most congested for their force of over 20,000 men so that they were densely packed with little space to ply weapons.

Henry ordered each archer to provide himself with a stake six feet in length and pointed at each end. In case of mounted attack, or when in position as on this very morning, the stakes were to be thrust into the ground, the upper ends sloping towards the enemy.

Provoking the French to attack them, the archers poured showers of arrows into the masses of cavalry and infantry who were laboriously struggling across the waterlogged field. When the cavalry were thrown back many riderless and unmanageable horses caused great disorder among the advancing men-at-arms, whose plodding progress was further disorganised by their being crowded in on each other owing to the 'funnelling' of the woods as they came nearer to the English lines. Goaded by the showers of arrows they tended to flinch away from the wedges of bowmen towards the three divisions of men-at-arms; so causing further loss of space. Even so, the English line was shaken by the first impact of the heavily armoured formations and a wild, mingled mass of desperately fighting men of all arms filled the area. Dropping their bows the archers seized their swords and axes and flung themselves into gaps or breaches. Almost completely exhausted by their struggle through the mud and the subsequent fighting, the French men-at-arms were no match for their lighter opponents; almost helplessly they stood until beaten to the ground by repeated blows from the archers' weapons. In a short time the French line was thrown back, reeling upon their second 'battle' as they plodded up to join the action.

On all sides the French were being rapidly despatched and the battle gradually petered out as fewer and fewer Frenchmen remained alive. For them it had been a nightmare. Pressing forward determinedly into the fight, they found it impossible to obtain sufficient room to wield their weapons; a man would be brought crashing to the muddy ground, taking with him those on either side of him, for all to wallow helplessly until despatched by the long, keen daggers of the archers. Defensive armour had become so heavy that there was no getting up once a man was down; in fact, the weighty men falling upon each other frequently caused death by suffocation.

Still fresh and in good order and so far uncommitted, the French third line might well have restored the failing fortunes of the day. But they stood, in indecision and dismay, realising that to advance was fruitless and not knowing what else to do, whilst on all sides the faint-hearted slipped away in the confusion.

The Battle of Verneuil (1424), said to have been harder than Agincourt, was an English victory after a near-disaster. As usual, each archer carried his double-pointed stake which he attempted to plant in the ground at about 250 yards' distance from the enemy. But the ground was hard so that the stakes would not plant easily and much time was wasted in trying to force them into the sunbaked earth; more time was consumed in passing the stakes forward from hand to hand until they reached the front rank. Before the hedge of stakes was half completed, the left-flank body of French mailed horse charged the archers and smashed through their ranks, forcing a passage over and through the half-erected stakes. The survivors of the ridden-down archers gathered together, forming a close-knit body for mutual protection, but the horsemen did not delay to deal with them, surging on towards the baggage-leaguer, where they were thrown back by archers. Although outnumbered 2–1 by French and Scots, the men-at-arms won the day after hand fighting.

At Rouvray in 1428, an English column of wagons formed itself into a laager on being attacked by a superior French force. But the French possessed artillery and they sat-off, bombarding the massed wagons. It seemed as though, for the first time in military history, guns alone were going to bring victory in the field, ushering in a new era of warfare.

After beating off an impetuous attack by the Scotch allies of the French, veteran Fastolf, the English commander, counter-attacked. Mounting his men on the baggage horses he sent them pouring from the twin openings in the leaguer, to fall upon the already shaken enemy. The conflict was short but sharp, and the Franco–Scottish force was soon routed and fleeing from the field.

Rouvray was not really a battle, but a small affair in which the English proved, as they had done in the greater fields of the past, that they were superior to the French in more ways than one. Not the least of these was the fact that the archers, the ordinary soldiers, were respected by the lords and knights who led them into battle. It was a very different situation from that prevailing in France, where the nobles arrogated to themselves alone the honour of bearing arms, despising the common soldier so that even in the 15th century French infantry were composed of the most wretched class of people. This was illustrated in the manner in which they were charged and ridden down by their own lords and commanders on those occasions when some distinguished act had aroused jealousy or scorn. The French foot soldier had reason to feel that for him to exhibit any true military qualities was simply to imperil his own life.

The second quarter of the 15th century found France really facing up to the fact that the old, chivalrous methods of warfare paid no dividends; that some effective innovation had to be discovered to counter the English system. Through bitter necessity and hard experience, the professional officers of France—Xaintrailles, La Hire and Dunois, for example—devised a simple method of minimising the superiority of the English archers—when the English were found drawn up in a good defensive position, the French refused to attack. Perhaps for the first time, the French admitted that with little or no chance of beating an English army in such a position, then there was no point in thrusting forward large bodies of troops as a target to be riddled with English arrows. The French commanders knew that the longbow had the ability to keep heavier-armed men at a distance but once cavalry or men-at-arms got among the archers and their supporting men-at-arms, weight of numbers might well decide the resulting mêlée—and the French could usually put more men into the field.

Now came some French victories through attacking the English when they were on the march, in camp or in towns where it was impossible to rapidly form an order of battle on ground specifically chosen for its defensive qualities. This reversal in the monotonous run of victories inevitably led to a lessening of English confidence so that a little more caution was displayed, initiative became stifled and plans could not be made with the former certainty of success. Naturally, with success and the knowledge that the English were not quite so sure of themselves, the French began to seek and win conflicts where they were able to hit the English before they could form up defensively.

The English commanders, with the traditions of Crécy, Poitiers, Agincourt and a host of similar but smaller battles behind them, disliked taking the offensive. The French commanders held off and refused to attack them in their carefully chosen position where they stood prepared and the English forces when unprepared were always liable to a sudden onslaught. Patay in 1429 is a fair example of the sort of conflict that now took place when an English force of 3,000 under Talbot and Fastolf was overwhelmed by a headlong French cavalry charge before it could take up a defensive position. It was all over very quickly; Talbot and most of the other leaders were captured, but Fastolf managed to get away, leaving behind his baggage and guns.

In 1450, the English lost the Battle of Formigny because their commanders were unintelligently influenced by the tactics of Agincourt, coupled with the fact that the French no longer made gross blunders at every engagement. It was the last battle but one to be fought in the Hundred Years War—a small-scale engagement but one that decided the fate of all Normandy. Galled by a prolonged bombardment from two French culverins, the English archers charged out from a good defensive position and captured the guns. Unsupported in the mêlée that ensued, the archers were slowly forced back so masking the fire of their comrades still behind their stakes. The arrival of French reinforcements on the English flank and rear ended the battle. Four-fifths of the English force were killed in this major disaster to English arms. By the use of intelligent offensive tactics it was a battle that could have been won before it even began, and then won again halfway through its course.

For more than 100 years the tactical employment of the English archer had brought success and each battle can be said to have favourably influenced the battle that followed. Crécy had been won because of the experience gained at Halidon Hill; Agincourt was, in its turn, influenced by Crécy and Poitiers. But now, the side that had always won began to lose because their opponents no longer played the game to the heavily loaded English rules, and, just as the French had been continuously beaten through a rigid adherence to outmoded tactics, now English commanders were being defeated because they persisted in slavishly applying the defensive tactics of Edward III and Henry V. After nearly a century the French had discovered a method that would minimise the deadly longbow.

In spite of the length of the war, until now there were surprisingly little changes or developments in arms, armament or method of fighting, with the exceptions of armour, which changed from mail to plate, and artillery. Little was heard of artillery in field operations, but in siege operations the power of the cannon was steadily increasing, so that in the last years of the war it had a predominating effect in securing the surrender of defended towns and castles.

Even in the late 15th and early 16th centuries, the longbow still retained its supremacy over the arquebus and had yet some famous fields to win, notably that of Flodden in 1513, where the old manoeuvres of Falkirk were repeated by both parties, the pikemen of the lowlands once again being shot to pieces by the archers of Cheshire and Lancashire. As late as the reign of Edward VI we find Kett's Insurgents beating, by the rapidity of their archery fire, a corps of German hackbuteers whom the government had sent against them. Nor was the bow entirely extinct as a national weapon in the days of Queen Elizabeth.

## STYLE OF FIGHTING

### *The English*

If one is to learn anything from the varying styles of fighting of the numerous nations that make up the military history of the Ancient and Medieval world then it is that battles are won by the commander who employs the capabilities of combined arms to their maximum. Edward I, Edward III and Henry V accomplished this, either knowingly or unwittingly, by combining the manoeuvrability of English archers skilled in the use of the longbow with the defensive capabilities of heavily armoured men-at-arms plus the use of armoured cavalry to attack or counter-attack. It was a method that brought to defence a marked tactical superiority over offence with dismounted men-at-arms acting as the defensive base for the devastating fire-power of the English archers in a system which has caused infantry to remain as the primary element of ground combat forces since the time of Crécy.

The English soldier of the Hundred Years War (perhaps excepting its last decade) was a finely drilled, disciplined and controlled soldier, led by at least three men who deserve to be ranked as 'Above-average' commanders.

The formations used in battle are fully detailed in the text of this chapter.

### *The French*

The French cavalry of this period were perhaps the finest in Europe in the sense that they were the best-born, natural handlers of horses and weapons. Unfortunately they were atrociously led and displayed all the vices of the feudal system—lack of control, ill-discipline, impetuosity, disinclination to co-operate, etc., etc.

At Crécy, the French were strung out in a long straggling column without any reconnoitring screen. On 'bumping into' the English King Philip endeavoured to halt the disorderly mass and concentrated his forces, either for the attack or to have them in hand for an overwhelming assault at daybreak on the following day. The mercenary Genoese crossbowmen were the only ones with sufficient discipline to be gathered together and they were sent forward to the assault, only to be decimated by the English archers and then ridden down by impetuous knights so filled with pride and impatience that they could not wait to fling themselves forward.

Ten years later at Poitiers, the French dismounted their horsemen on the false supposition that this was the only way in which they could beat the similarly dismounted English men-at-arms. But in so doing, they were deprived of their principal asset for offensive action—mobility and shock. The French commander typically gave no thought to any kind of tactical manoeuvre that might turn the outnumbered English from their strong position.

Even seventy years later at Agincourt, the French leaders were still unable to control their impetuous and undisciplined nobility who it is said had 'forgotten nothing and learned nothing'. As a result, the battle went as the others had done when a vastly outnumbered English force of courageous and disciplined archers supporting dismounted men-at-arms in a defensive formation again shattered medieval heavy cavalry and infantry.

In the later stages of the Hundred Years War the French commanders, at last recognising the superiority of English archery fire-power, avoided attacks against the English in prepared positions and seized all possible opportunities to force them to fight at a disadvantage. Du Gueschlin, an outstanding Constable of France, excelled in night attacks and other such stratagems despite English protest that such practices were unchivalrous and 'Unknightly'. Fittingly, the final French successes of the Hundred Years War were based on the same doctrine that had brought success to the English—the superiority of good infantry on the defensive when well supported by missile-power. The French substituted artillery weapons for longbows and crossbows.

*The Scots*

The Scots were the first unfortunate soldiers to feel the full weight of the successful English military precept of heavy fire-power, strong defence and final heavy counter-attack. In rigid formations of pikemen known as schiltrons strongly resembling the phalanx of the Macedonians or the Swiss formation, the Scots fought bravely and well but, lacking missile-power of their own, were never in with a chance.

A single chart would be unfair to the French because of the vast disparity between their knightly cavalry and lowly infantry—hence a chart for each. Note that mercenary troops are NOT included.

Like the French, they took a long time to learn and were still fighting and being beaten in the same old fashion, both in Scotland and as Allies of the French, for nearly a century.

The Scottish soldier was not particularly well disciplined or controlled (he probably remained in his schiltron because instinctively he felt it was a strong and life-preserving formation). Occasionally, he was led by outstanding commanders and can be said to have been unfortunate in being a tough soldier defeated by a new system beyond his understanding and ability.

*Fighting Assessment Chart:* ENGLISH

| Factor | Rating | Points |
|---|---|---|
| 1. Ability to assault in controlled order | Reasonable | 2 |
| 2. Bravery | First Class | 3 |
| 3. Control | First Class | 3 |
| 4. Defensive ability | First Class | 3 |
| 5. Desperation | Not applicable | – |
| 6. Discipline | First Class | 3 |
| 7. Ferocity | Not applicable | – |
| 8. Manoeuvrability | Reasonable | 2 |
| 9. Mobility | Reasonable | 2 |
| 10. Movement, speed of | Reasonable | 2 |
| 11. Order | First Class | 3 |
| 12. Tactical Sense | First Class | 3 |
| 13. Training and Drilling | First Class | 3 |
| 14. United action, capability for | First Class | 3 |
| 15. Weapon Handling | First Class | 3 |
| 16. Demoralisation, likelihood of | Unlikely | 3 |
| 17. Impetuous Action, likelihood of | Rare | 2 |
| 18. Panic, likelihood of | Unlikely | 3 |
| 19. Reluctance to take definite action | Unlikely | 3 |
| 20. Unscheduled Manoeuvres, likelihood of | Rare | 2 |
| | *Fighting Assessment* | 48 |

*Fighting Assessment Chart:* FRENCH INFANTRY

| *Factor* | *Rating* | *Points* |
|---|---|---|
| 1. Ability to assault in controlled order | None | 0 |
| 2. Bravery | Average | 1 |
| 3. Control | None | 0 |
| 4. Defensive ability | None | 0 |
| 5. Desperation | Not applicable | – |
| 6. Discipline | Average | 1 |
| 7. Ferocity | Not applicable | – |
| 8. Manoeuvrability | Average | 1 |
| 9. Mobility | Average | 1 |
| 10. Movement, speed of | Average | 1 |
| 11. Order | None | 0 |
| 12. Tactical Sense | None | 0 |
| 13. Training and Drilling | Average | 1 |
| 14. United action, capability for | None | 0 |
| 15. Weapon Handling | Average | 1 |
| 16. Demoralisation, likelihood of | Likely | 0 |
| 17. Impetuous Action, likelihood of | Likely | 0 |
| 18. Panic, likelihood of | Likely | 0 |
| 19. Reluctance to take definite action | Likely | 0 |
| 20. Unscheduled Manoeuvres, likelihood of | Likely | 0 |
| | *Fighting Assessment* | 7 |

*Fighting Assessment Chart:* FRENCH CAVALRY

| | *Factor* | *Rating* | *Points* |
|---|---|---|---|
| 1. | Ability to assault in controlled order | None | 0 |
| 2. | Bravery | First Class | 3 |
| 3. | Control | None | 0 |
| 4. | Defensive ability | Little | 1 |
| 5. | Desperation | Not applicable | – |
| 6. | Discipline | Little | 1 |
| 7. | Ferocity | Little | 1 |
| 8. | Manoeuvrability | Little | 1 |
| 9. | Mobility | Reasonable | 2 |
| 10. | Movement, speed of | Reasonable | 2 |
| 11. | Order | None | 0 |
| 12. | Tactical Sense | None | 0 |
| 13. | Training and Drilling | Good | 3 |
| 14. | United action, capability for | Little | 1 |
| 15. | Weapon Handling | Good | 3 |
| 16. | Demoralisation, likelihood of | Average | 1 |
| 17. | Impetuous Action, likelihood of | Likely | 0 |
| 18. | Panic, likelihood of | Average | 1 |
| 19. | Reluctance to take definite action | No | 3 |
| 20. | Unscheduled Manoeuvres, likelihood of | Likely | 0 |
| | | *Fighting Assessment* | 23 |

*Fighting Assessment Chart:* SCOTS

| | *Factor* | *Rating* | *Points* |
|---|---|---|---|
| 1. | Ability to assault in controlled order | Reasonable | 2 |
| 2. | Bravery | First Class | 3 |
| 3. | Control | Average | 1 |
| 4. | Defensive ability | First Class | 3 |
| 5. | Desperation | Not applicable | – |
| 6. | Discipline | Average | 1 |
| 7. | Ferocity | Not applicable | – |
| 8. | Manoeuvrability | Average | 1 |
| 9. | Mobility | Average | 1 |
| 10. | Movement, speed of | Average | 1 |
| 11. | Order | Average | 1 |
| 12. | Tactical Sense | Average | 1 |
| 13. | Training and Drilling | Average | 1 |
| 14. | United action, capability for | Average | 1 |
| 15. | Weapon Handling | Reasonable | 2 |
| 16. | Demoralisation, likelihood of | Average | 1 |
| 17. | Impetuous Action, likelihood of | Average | 1 |
| 18. | Panic, likelihood of | Average | 1 |
| 19. | Reluctance to take definite action | Average | 1 |
| 20. | Unscheduled Manoeuvres, likelihood of | Likely | 0 |
| | | *Fighting Assessment* | 23 |

# THE MERCENARIES

From the earliest times professional soldiers have thrust aside nationalistic and racial feelings, forgotten customs or language and sold their services to the highest bidder. A trained and experienced soldier, with few scruples about the use to which the weapon he had to trade was put, could always depend upon being well-paid (in cash or plunder) in return for his skilful services at the risk of his life and limbs.

In the ancient world, feudalism or some similar system transformed peasants into serfs, yet many of them were able to become relatively prosperous with a degree of independence—in either case it often meant that they lacked the heart and incentive to fight. Then the state or kingdom had no recourse but to hire foreign mercenaries, professional soldiers of superior fighting ability with nothing more than financial attachment to their employers. Becoming indispensable as the citizens became less and less accustomed to bearing arms, mercenaries made increasing demands for more pay and greater privilege. In time this blackmailing failed to come up to their expectations and, conscious of their strength, they overthrew those who hired them and set themselves up in their place. This meant that any state who used mercenary troops had to constantly be on guard against the very forces which had been employed to protect it.

Sometimes mercenaries were more trouble than they were worth. With the coming of peace and their employment ended, these private armies turned into gangs of bandits living off the countryside. One of the causes of King John's unpopularity was that he brought over to England the mercenary crossbowmen of Fawkes de Breaute, ruffians who were still around in the next reign. It is considered likely that the highly organised mercenary bands of the 15th century deliberately dragged out the wars in which they were employed since complete victory for either army would mean peace and an end to pay as they were disbanded. Their leaders were veteran soldiers who believed in the principle of 'if you scratch my back I'll scratch yours'. Mercenaries of other bands who fell into their hands could always rely on being released for a moderate ransom. Conscious that next year it might come to pass that they would have a change of master and become soldiers of the king whom they had defeated in the previous year (providing he made them a better offer than their present master) they never killed the goose that laid the golden egg by butchering a rival army or laying waste to a defeated town.

Since the beginnings of warfare, history abounds with recorded instances of mercenary soldiers fighting, often far away from their native lands—for generations Northmen went down to Constantinople and enlisted in the Varangian Guard, which kept the Byzantine Emperor on his tottering throne. As early as 1288 B.C. Ramses had a large army which included mercenaries who fought in the Battle of Kadesh. After the Pelponnesian War ended in 404 B.C., many Greeks had no other skill than that of arms so they became professional soldiers. There had been Greek mercenaries before but now their numbers increased. In 401 B.C. Cyrus hired 10,000 Greeks to fight for him in revolt against his brother Artaxerxes, King of Persia. Between 399 and 375 B.C. 25,000 Greeks took service as mercenaries abroad and many of them brought back the techniques they learned from such as the Rhodan slingers and the Cretan archers. Shortly after 400 B.C. in Athens, Iphicrats raised a body of mercenary javelinmen called peltasts modelled on the Thracians, who in all but big set battles were the most effective infantry of their time.

In the last years of his wars Alexander relied increasingly on mercenaries to guard his communicatons and raised a force of horsemen from the nomads of eastern Iran. At the battle of Gaugamela, the Persian King Darius had 2,000 Greek mercenary Hoplites fighting against Alexander who had also recruited a number of Greek mercenaries, including Cleander's veteran mercenary cavalry and mercenary Greek hoplites. The army of Hannibal consisted mostly of mercenaries held together by Hannibal's leadership and the prospect of plunder. Probably the best of them were the Numidian mounted javelinmen under a brilliant cavalry commander, Maharbal. The development of mercenary armies in Rome produced military successes abroad but brought internal policies controlled by force of arms. Stilicho, a very able Vandal general, commanded the western Roman armies that drove Alaric's Visigoths out of Greece. The Emperor Honorius, fearing his power, had Stilicho murdered so provoking the 30,000 Gothic mercenaries in the Imperial Army to mutiny.

Attila the Hun hired out his ferocious riders as mercenaries; 6,000 of them were with Belisarius when he set out from Byzantine in 532 with a multi-national mercenary army of 10,000 infantry and 5,000 cavalry including Isaurian mountaineers and 4,000 Heruls.

In 1066, William of Normandy built up his army for the invasion of England by enticing volunteers and mercenaries from Normandy from all parts of France and even from the Norman adventurers of southern Italy. After the defeat of Harold's Saxons, the Normans took over the whole of England, later extending their conquests into Ireland when the Norman 'Strongbow' transported South Welsh longbowmen with him to fight the peasants in their peat-bogs. In the early 11th century A.D. Norman knights and their followers ranged the Mediterranean as freebooting mercenaries and it is recorded that Robert, son of Tancred, Baron of Hauteville, was making a good living in 1046 as a part-time mercenary, part-time brigand and cattle thief, operating from the mountains of Calabria in southern Italy. Whilst the backbone of the Crusading armies was the knights, feudal tenants and members of the religious Orders, they were increasingly supplemented by mercenaries and 'Turcopoles' (renegade Moslems).

These were days when feudal levies were indisciplined and often it was impossible to raise enough of them to make up an army, so mercenaries were employed because they knew their job and were obedient as long as they were paid. In the Civil Wars of 1215–1217, which racked England after the breakdown of the

Runnymede agreement, Gascon crossbowmen were hired by King John—they were efficient professionals who could be hired and fired as the occasion demanded. Later, in England the King allowed his tenants to commute their feudal service by payment of a tax (scutage). With the funds so derived he hired mercenary adventurers, trained and experienced professionals accustomed to working together who would serve their paymaster as long as they were required. Thus, in the 13th century began an increasing dependence on mercenaries. But Edward I found in his Welsh and Scottish Wars that Gascon crossbowmen were very expensive to hire—in 1277 it is recorded that an arbalest cost the equivalent of twenty-five pounds and each bolt shot from it a shilling or one shilling and sixpence.

As early as the 13th century men of Brabant and Flanders, skilled Brabancon pikemen, formed themselves into mercenary bands and hired themselves out in France and Italy before the crossbow put them out of business. This was the weapon of the men who became the chief mercenary infantry of Europe—the crossbowmen of the northern Italian cities, principally Genoa. Their weapon was heavy, slow to operate and unable to work in wet weather but it was accurate up to a range of about 100 yards, it could be handled by relatively untrained men who were not strong enough to draw the longbow and it was a particularly good weapon for defensive loophole shooting. The Genoese were well protected by steel cap and mail shirt. They were probably the best troops in the French army at Crécy in 1346 but their crossbows suffered from a sudden shower of rain that immediately preceded the battle so that they were relatively ineffective and they were ridden down by the French knights, impatient to get to grips with the English array. This was characteristic of these men—in 1311 the proud French knights of the Duke of Athens were lured into a swamp in Greece by the mercenary infantry of the Catalan Grand Companies and perished almost to a man. They were led by the first of the great condottieri leaders, Roger Flor, a renegade Templar, who gathered and combined all the paid-up mercenaries of Spain, and with this Catalan army marched as far as the Levant, bringing the Byzantine Emperor Andronicus to his knees in 1308, so carving out for themselves a permanent home in the Duchy of Athens.

Early in the Hundred Years War, Edward III had a few bands of foreign mercenaries serving on indentures under highly capable professional officers. Warfare in the Middle Ages became commercialised as standing armies and mercenary troops replaced the feudal levy and its unsatisfactory short, forty-day service. Clad now in plate-armour and thus more than ever an expensive item, many a man-at-arms found he could sell his services substituting loyalty to his lord by allegiance to the highest bidder. This was often the King, who found the mercenary a better prop to his power than the fractious vassal.

Sometimes mercenaries sold themselves as individuals but more often they banded together into small (and sometimes not particularly well organised) bands under a chosen leader; calling themselves such picturesque names as 'The White Company'; 'The Company of the Star' or '. . . of the Rose'. By the end of the 13th century a condottiere system was established as noted mercenary chiefs collected large groups of soldiers and hawked their services from court to court throughout Europe. One such captain was the Englishman Sir John Hawkwood who was highly priced because he insisted on his 'White Company' always giving their very best to his current master. Other captains included 'Arch Priest' Arnold of Cervole, Fra Moriale and his 'Grand Company'—a captain who had been expelled from the Order of Saint John for his less than religious attitude to war. Naturally these professional soldiers experimented with the weapons which came closest to their hands. Moriale used 7,000 mounted men-at-arms to 2,000 auxiliary crossbowmen; while Hawkwood, being English and recalling Crécy, employed 2,500 cavalry and 2,000 longbowmen. In their early days, the condottiere favoured horses because often they were knights who had tired of the lists and had forsaken the boredom of peace to follow a warlike career. Many of them were landless younger sons or adventurers, but mostly they were professional soldiers who knew no other occupation than the pursuit of arms in which they made themselves expert. In time, battles began more and more to be fought by these professionals who worked hard during the campaigning season of March to October before resting in winter quarters or nursing themselves back to health after a wound.

The condottiere of the later Middle Ages who waged the endless little 15th century wars between the Italian city-states were often knights or horsemen in heavy plate-armour or else high-priced Swiss mercenary pikemen ('Point d'Argent, Point de Suisse' was a saying of the time) or landsknechte (mercenary German pikemen formed on the Swiss model). These professionals devised a peculiar and highly personalised system of warfare designed to reach a decision without excessive loss or personal danger to themselves so that, for example, the cavalrymen could decide a conflict by a dashing charge which often routed the enemy without many blows being struck. It was a system that was only possible if pitched battles were avoided and forces operated over flat and dry country not cut up by ditches so permitting the horsemen to manoeuvre and deploy under spacious and unimpeded conditions. This meant that battles between bands of condottiere was largely a matter of extensive manoeuvring, concealment and other ruses, to outflank or encircle the enemy, or to force him into a position from which he could not attack with any advantage against your own position on hard ground or roads. The Swiss were notable exceptions to this system, being ferocious and merciless.

But the main object was to win with as few men killed as possible because a trained mercenary was a valuable commodity whose death represented to his leader both the loss of a skilled soldier and a financial investment.

## STYLE OF FIGHTING

Mercenary units usually formed part of an army and their particular style of fighting was presumably considered in the commander's battle plan. This meant that mercenaries fought according to their method, type and style—so that Greek hoplites fought as heavy infantry, Cretan slingers and Genoese were used as light troops for skirmishing and missile-power whilst Swiss and landsknechte fought in their customary formation of massed pikes. Many soldiers who had been in forces that employed mercenaries, returned to their homeland bringing with them a considerable knowledge of unusual weapons used by mercenaries and the manner in which they were handled.

Being professional soldiers, mercenaries were accustomed to being disciplined and controlled to obeying orders besides being masters of the personal weapons they used. All this made them worth having on your side, but on the other hand, they rarely had any sense of loyalty except perhaps to their own immediate commander and certainly not to the overall commander of the army or to the king or prince who employed them. Military history abounds with instances of perfidious hired soldiers—here are a few examples.

After the death of Alexander his generals wrangled amongst each other over his Empire in a struggle notable for its cunning, treachery and bribery. It is said that their armies were completely mercenary and would turn against any leader if offered more money by a rival. Hamilcar Barca, Carthaginian hero of the First Punic War, had to handle a revolt of 25,000 mercenaries who laid siege to Carthage in 238 B.C. By a brilliant stratagem he got 10,000 men out of the city and defeated the rebels at the Battle of Utica and soon after ambushed and defeated another army of rebellious mercenaries near Tunis. In 378 A.D. the Gothic forces who defeated the Romans at the Battle of Adrianople were assisted by a force of Gothic mercenaries who deserted from the Roman army. In their struggles against the barbarians during the middle of the 5th century A.D., a barbarian General Odoacer led a mutiny of most of the barbarian mercenaries in the Roman General Orestes's army and defeated him at the Battle of Pavia in 476 when Orestes was captured and killed. In 1069, Romanus, an able Byzantine general, helped to subdue a mutiny among his mercenaries and two years later lost the Battle of Manzikert when Kipchak deserted with his Pecheneg Light Cavalry, so bringing the strength of Romanus's army down to 35,000 men facing far greater numbers of Turks. This defeat deprived the Byzantine Empire of the principal native recruiting ground for the Empire so that for the next four centuries the Byzantine Empire was forced to rely on mercenaries for the bulk of its armed forces.

So it can be seen that loyalty and treachery must be considered when employing mercenaries so that the wargamer would do well to employ a 'Treachery' factor.

In the Middle Ages feudal armies, with levies who could only be called to action for a few weeks of the year, could rarely be maintained long enough to undertake the prolonged sieges necessary to capture the powerful forts that existed. The alternative was mercenary armies but these were extremely expensive to maintain indefinitely in the field. Thus, when wargaming in a campaign sense, a reasonable strategy would be to drag the heels so that feudal armies dispersed with the passing of time whilst mercenary forces became too expensive to maintain in the field.

Swiss pikemen became the leading mercenary troops in Europe. They had a reputation for refusing to fight if unpaid, nor would they take up arms against a force also employing Swiss mercenaries. Subsequently Maximilian of Hapsburg created a German counterpart, the landsknechte, to offset the Swiss infantry advances. Wherever these two came up against each other the struggle was fierce and bloody and the Swiss were invariably victorious, being even more ferocious and bloodthirsty against their German imitators than against any other foe. The employers of these mercenary pikemen found that the greatest value was obtained from them by fitting them into a combined-arms team with heavy and light cavalry.

When Swiss mercenaries are employed, it is advisable to build into the framework of the rules a 'Ferocity' factor so arousing a sense of terror in their opponents before the battle commences.

The last of the mercenaries that comes within the scope of the period of this book are perhaps the least worthy of them all. Prominent in the 15th century when the numerous small States of Italy were ceaselessly and senselessly battling with each other, the mercenary condottierie were always for sale to the highest bidder right up to the eve of battle and sometimes during the actual fighting itself. From their commander down to their lowest ranks, they had no personal interest in patriotism or victory except as a factor in lining their pockets. Most of the bands were heavy cavalry, the safest and most comfortable way of fighting in that day because they were so thoroughly protected by their armour that bloodshed and death became almost rare. Because of the mountains of Italy or the extensive marshlands around the valley of the River Po, heavy cavalry were the least suited of all troops to be employed in the area so that battles became almost like pre-arranged football matches, held in selected areas. The engagements were fought in the manner of a chess game and hastily concluded when one army found itself in such an unfavourable position that it as unlikely to win, when it would concede victory and, without heavy losses or perhaps any losses at all, they ransomed their prisoners and marched away. This paradise for the professional soldier collapsed when Charles VIII of France brought a well trained and hard fighting army into Italy towards the close of the 15th century.

This latter period of military history, sterile as it may have been in real life, provides two interesting wargaming derivations in the 'Honours of War' system and the 'Manoeuvre' game. (Both are explained in detail elsewhere in this book.)

*Fighting Assessment Chart:* LANDSKNECHTE

| | *Factor* | *Rating* | *Points* |
|---|---|---|---|
| 1. | Ability to assault in controlled order | First Class | 3 |
| 2. | Bravery | Reasonable | 2 |
| 3. | Control | First Class | 3 |
| 4. | Defensive ability | Reasonable | 2 |
| 5. | Desperation | Against Swiss Only | 3 |
| 6. | Discipline | Reasonable | 2 |
| 7. | Ferocity | Not applicable | – |
| 8. | Manoeuvrability | First Class | 3 |
| 9. | Mobility | First Class | 3 |
| 10. | Movement, speed of | First Class | 3 |
| 11. | Order | First Class | 3 |
| 12. | Tactical Sense | Reasonable | 2 |
| 13. | Training and Drilling | First Class | 3 |
| 14. | United action, capability for | First Class | 3 |
| 15. | Weapon Handling | First Class | 3 |
| 16. | Demoralisation, likelihood of | Rare | 2 |
| 17. | Impetuous Action, likelihood of | Unlikely | 3 |
| 18. | Panic, likelihood of | Rare | 2 |
| 19. | Reluctance to take definite action | Rare | 2 |
| 20. | Unscheduled Manoeuvres, likelihood of | Unlikely | 3 |
| | | *Fighting Assessment* | 50 |

# APPENDIX 1

# Availability of Wargames Figures

This chart lists figures suitable for wargaming that are available from the named commercial manufacturers. The 'flats' section has been compiled by J. R. Granville Bantock and Dr Alastair Bantock. In addition to the listed makers it is known that figures for this period are also made by:

Alymers Miniploms (of Spain)
Elastolin (of W. Germany)
The Garrison (W. and P. Militaria of South Harrow)
Hinton Hunt Figures (of Taplow)

For addresses and further information on all named manufacturers (and others) see *Handbook for Model Soldier Collectors* by Donald Featherstone (Kaye and Ward, London, 1969).

The ever-increasing series of 20 (25) mm plastic figures manufactured by Airfix possess the merit of being readily convertible to *any* desired soldier of any period. This invaluable fact puts wargaming in the Ancient and Medieval period within the reach of all. For instruction in converting Airfix figures see chapter 5, *Military Modelling* by Donald Featherstone (Kaye and Ward, London, 1970) and regular series in *Wargamer's Newsletter* and *Airfix Magazine*. Of particular interest and value to wargamers specialising in the Ancient period is a series of articles 'Roman Friends and Foes' by Bob O'Brien, published in *Airfix Magazine* for February 1968 and November/December 1968; January, February, March, April, May and June 1969. This series not only details the general characteristics and tactics of both Romans and their numerous enemies, but also gives explicit instructions on converting standard Airfix figures into numerous other types of warriors, such as the Goths, the Picts, the Ancient Britons, Numidians, Gauls, Germans and Arabs, besides dealing with specific types such as camel and armoured horse, archers, slingers, Roman artillery and siege weapons, war elephants, chariots, boats, etc. The series is well illustrated with photographs and drawings.

| *Type of Soldier* | *Available From* *Solids* | *Flats* | *Figures available that can substitute or readily convert* |
|---|---|---|---|
| Sumerians | | Neckel<br>Blum<br>(Gottstein) | |
| Ancient Egyptians | Miniature Figurines<br>Scruby | Ochel<br>Rössner<br>Vollrath<br>Neckel<br>Hafer<br>Bittner<br>Emmersling<br>Lecke | |
| Hyksos | | Neckel | |
| Hittites | Scruby | Blum | Miniature Figurines |
| Assyrians | Miniature Figurines<br>Scruby | Ochel<br>Blum<br>Neckel | |
| Medes | Miniature Figurines<br>Scruby | Neckel<br>Hafer | |
| Greeks | Miniature Figurines<br>Scruby | Ochel<br>Retler<br>Neckel<br>Manz | |
| Trojans | Miniature Figurines<br>Scruby | Retler | |
| Athenians | Miniature Figurines<br>Scruby | Ochel<br>Neckel<br>Retler | |
| Persians | Miniature Figurines<br>Scruby | Ochel<br>Retler<br>Cortum | |
| Spartans | Miniature Figurines<br>Scruby | Ochel<br>Neckel<br>Retler | |

| *Type of Soldier* | *Available From* *Solids* | *Flats* | *Figures available that can substitute or readily convert* |
|---|---|---|---|
| Thebans | Scruby | Ochel<br>Neckel<br>Retler | Miniature Figurines |
| Macedonians | Miniature Figurines<br>Scruby | Ochel<br>Neckel<br>Cortum<br>Rössner | |
| Romans (various periods) | Miniature Figurines<br>Scruby<br>Airfix | Ochel<br>Vorburg<br>Scholtz<br>Tobinnus<br>Neckel<br>Cortum<br>Mignot<br>Manz<br>Bittner | |
| Roman Auxiliaries | Miniature Figurines<br>Scruby<br>Airfix | Ochel<br>Vorburg<br>Scholtz<br>Tobinnus<br>Neckel<br>Cortum<br>Mignot<br>Manz<br>Bittner | |
| Carthaginians | Miniature Figurines<br>Scruby | Ochel<br>Blum (Gottstein)<br>Scholtz | |
| Gauls | Miniature Figurines<br>Scruby | Ochel<br>Tobinnus<br>Mignot<br>Scholtz | Airfix |
| Parthians | Miniature Figurines | Vollrath | |
| Goths | Miniature Figurines | Pohl | |
| Vandals | | Pohl | Miniature Figurines |

| *Type of Soldier* | *Available From* *Solids* | *Flats* | *Figures available that can substitute or readily convert* |
|---|---|---|---|
| Britons | Miniature Figurines<br>Airfix | Blum | |
| Barbarians (cavalry) | Miniature Figurines<br>Scruby | Beck<br>Ochel | |
| Huns | Miniature Figurines<br>Scruby | Pohl | |
| Mongols | Miniature Figurines<br>Scruby | Pohl | |
| Visigoths | | Pohl | |
| Saxons (fyrd; thegns; huscarls) | Miniature Figurines | Blum | Airfix |
| Danes | Miniature Figurines | Pohl<br>Beck | Airfix |
| Vikings | Miniature Figurines | Pohl<br>Beck | Airfix |
| Normans | Miniature Figurines | Blum<br>Muller | |
| Byzantines | Miniature Figurines<br>Scruby | Vollrath | |
| Seljuk Turks | | Muller | |
| Varangarian Guard | Miniature Figurines | Heinrichson | |
| Franks | Miniature Figurines<br>Scruby | Blum (Gottstein) | |
| Moors | Miniature Figurines | Blum (Gottstein) | Airfix |
| Chinese | Miniature Figurines | Hafer<br>Emmersling<br>(Thro' Donath)<br>Grosch | |
| Japan (incl.) Samurai) | | Rössner<br>(Thro' Ochel) | |

| *Type of Soldier* | *Available From* *Solids* | *Flats* | *Figures available that can substitute or readily convert* |
|---|---|---|---|
| India (Hindu) | | Hafer<br>Retler | |
| Crusaders | Miniature Figurines<br>Scruby | Muller<br>Maier<br>E. Ortmann<br>Manz | |
| Saracens | Miniature Figurines<br>Scruby | Muller<br>Maier<br>Manz | |
| Hospitallers & Knights Templar | Scruby | Muller | |
| Knights of the Teutonic Order | Miniature Figurines<br>Scruby | Hafer<br>Muller<br>H. Bittner<br>Neumeister<br>Manz | Scruby |
| Ottoman Turks | Miniature Figurines | Muller | |
| Hungarians | | Muller<br>Ortmann | |
| Gascon Cross-bowmen | Scruby | Muller | |
| Turks (Spahis; Janissaries, Bashi-Bazouks, etc.) | Miniature Figurines | Muller<br>Fohler<br>Neckel | |
| Swiss | Miniature Figurines<br>Scruby | Tobinnus<br>Blum (Gottstein) | |
| Austrians (14th Century) | Scruby | Ochel | |
| Brabancons (pikemen) | | | Scruby |
| Flemish (pikemen) | | | Scruby |
| Burgundians | | Blum | Scruby |

| *Type of Soldier* | *Available From* *Solids* | *Flats* | *Figures available that can substitute or readily convert* |
|---|---|---|---|
| Landsknechte | Miniature Figurines | Tobinnus<br>Ochel<br>Lecke<br>Kebbel<br>Blum<br>Neckel<br>Manz<br>Münchow | |
| Spanish (15th/16th century) | Miniature Figurines | Tobinnus | |
| Scots (14th century) | Miniature Figurines | Muller | |
| French (14th/15th centuries) | Miniature Figurines | Rössner<br>Muller<br>Ochel | |
| English (14th/15th centuries) (incl. knights; men-at-arms; longbowmen; crossbowmen; etc.) | Miniature Figurines<br>Airfix | Manz<br>Muller<br>Ochel | Airfix |
| Genoese crossbowmen | Miniature Figurines | Muller | Scruby |

## APPENDIX 2

# Pictorial Sources for Ancient and Medieval Troops

The following books contain illustrations, some coloured, of the soldiers who formed the armies considered in this volume:

Courtland Canby, *A History of Weaponry* (Hawthorns Book Inc., New York, 1963)
Jack Coggins, *The Fighting Man* (Doubleday, New York, 1966)
Henry Treece and Ewart Oakeshott, *Fighting Men* (Brockhampton Press, Leicester, 1963)
S. E. Ellacott, *Spearmen to Minuteman* (Abelard–Shuman, 1965)
D. Halevy, *Armies and their Arms* (Odhams Press, 1967)
Karel Toman, *A Book of Military Uniforms and Weapons* (Paul Hamlyn and Allan Wingate, 1964)
Liliane and Fred Funcken, *Le Costume et les Armes des Soldats de tous les Temps* (Casterman, Belgium, 1966)

This chart lists page numbers in the relevant books.

| *Troops* | *Canby* (*page*) | *Coggins* (*page*) | *Treece* (*page*) | *Ellacott* (*page*) | *Halevy* (*page*) | *Toman* (*page*) | *Funcken* (*page*) |
|---|---|---|---|---|---|---|---|
| Sumerians | 10 | | | | | | |
| Ancient Egyptians | 10 | 4 | 19 | 12<br>18 | 9–11 | 8–11 | 11–15 |
| Assyrians | 10 | 6–11 | 23 | | 12–13 | 13–17 | 15–23 |
| Greeks | 13 | 16–21 | 28–9 | 25 | 15 | 20–2 | 30–2 |
| Spartans | | | 33 | | | | 45<br>51 |
| Persians | | 13–15 | | | | 18–20 | 25–7 |
| Macedonians | | | | | 16 | 24–5 | 43–5<br>49 |
| Romans | | 48–9<br>57<br>65<br>74<br>78–9 | 45<br>49<br>55 | | 20 | 28–31 | 57–69 |
| Carthaginians | | | | | 23 | 26 | |
| Ancient British | | | 10<br>35 | | | | 75–9 |
| Huns | | | | | 24 | 36 | 83 |
| Goths | | | 75 | | | 35–7 | |
| Mongols | 32 | 122 | | | | | |
| Germans | | | | 48 | 26 | | 81 |

| *Troops* | *Canby* (*page*) | *Coggins* (*page*) | *Treece* (*page*) | *Ellacott* (*page*) | *Halevy* (*page*) | *Toman* (*page*) | *Funcken* (*page*) |
|---|---|---|---|---|---|---|---|
| Gauls | | | | | 22 | 34 | 75 |
| Saxons | | 108 | | | | | 87<br>93 |
| Vikings | | 88–92 | 92 | 64 | | 40 | 91 |
| Normans | | 108–11 | 111 | 69 | 29 | 44–5 | 95 |
| Byzantines | | 103 | | | | 53 | |
| Franks | | | | 67 | 26–7 | 39 | 85–8 |
| Chinese | 28–9 | 312 | | | | | |
| Japanese | 30–7 | 291–8 | | | 35 | | |
| Indians | 32–3 | | | | | | |
| Crusaders | | 114–16 | | | 31 | 46–7 | 95 |
| Saracens | | 114–16 | | | | 47 | 91 |
| Ottoman Turks | | | 176–7 | | 31 | | 117 (6–9) |
| Swiss | 58–9 | 141–6 | 145–9<br>176–7 | | 41 | 71–2<br>68–9 | |
| Hundred Years' War | | 131–7 | 145 | 85 | 32 | 58–60 | 105–13 |
| The Mercenaries | | | | 97 | | 66–7 | |
| Advent of Gunpowder | | | 191 | 98 | 36 | 76 | 115 |

In addition many of the books and articles listed in the specific bibliography for each period include illustrations of soldiers, their arms and equipment.

## APPENDIX 3

# Suggested Reading

A vast period of military history has been but briefly explored in this volume. The books listed here cover that period in detail, both in a general sense and specifically.

### GENERAL


Birnie, A., *The Art of War* (1942)
Boudet, J. (Ed.), *The Ancient Art of Warfare* (1969)
Bowood, R., *Soldiers Soldiers* (1965)
Canby, Courtland, *A History of Weaponry* (1963)
Chandler, D., *Battlefields of Europe* (1965)
Coggins, J., *The Fighting Man* (1966)
Creasy, E. S., *The 15 Decisive Battles of the World* (1908)
Denison, G. T., *A History of Cavalry* (1913)
Eggenberger, D., *A Dictionary of Battles* (1967)
Ellacott, S. E., *Spearmen to Minutemen* (1965)
Falls, C., *The Art of War* (1961)
Featherstone, D., *Wargames* (1962)
—— *Advanced War Games* (1969)
—— *War Game Campaigns* (1970)
—— *Battles with Model Soldiers* (1970)
Fuller, J. F. C., *Decisive Battles of the Western World*, vol. I (1939)
Funcken, L. and F., *Le Costume et les Armes des Soldats de tous les Temps*, vol. I (1966)
Halevy, D., *Armies and their Arms* (1962)
Hall, D. G. E., *A History of S.E. Asia* (1955)
Hazard, H. W., *Atlas of Islamic History* (1954)
Laffin, J., *Links of Leadership* (1966)
Latourette, K. S., *The Chinese, Their History and Culture* (1934)
Liddell-Hart, B. H., *Great Captains Unveiled* (1927)
—— *The Decisive Wars of History* (1929)
Lloyd, E. M., *A Review of the History of Infantry* (1908)
Montgomery of Alamein, *A History of Warfare* (1968)
Montrose, L., *War Through the Ages* (1944)
Oakshott, R. E., *The Archaeology of Weapons* (1960)

Oman, C. W. C., *The Art of War in the Middle Ages, A.D. 378–1515* (1953)
Preston, R. A., Wise, S. F., and Werner, H. O., *Men in Arms* (1956)
Spaulding, O., Nickerson, H., and Wright, John, *Warfare: A Study of Military Methods from the Earliest Times* (1937)
Starr, Chester G., *A History of the Ancient World* (1965)
Sun Tzu (trans. Griffith, S. B.), *The Art of War* (1963)
Toman, K., *A Book of Military Uniforms and Weapons* (1964)
Treece, H., and Oakshott, E., *Fighting Men* (1963)
Turney-High, H. H., *Primitive War* (1949)
Weller, J., *Weapons and Tactics* (1966)
Whitman, J. E. A., *How Wars are Fought* (1944)

### WEAPONS

Burke, E., *The History of Archery* (1958)
Featherstone, D., *The Bowmen of England* (1967)
Payne-Gallway, Sir Ralph, *The Crossbow, Medieval and Modern, Military and Sporting* (1958)

### BIBLICAL PERIOD

Belloc, H., *The Battleground—Syria* (1936)
Cottrell, L., *Life Under the Pharaohs* (1955)
Gale, Sir Richard, *Great Battles of Biblical History* (1967)
Gurney, O. R., *The Hittites* (1952)
Yadin, Yigael, *The Art of Warfare in Biblical Lands* (1963)

### THE GREEKS

Adcock, F. E., *The Greek and Macedonian Art of War* (1957)
Allcroft, A. H., *Peloponnesian War* (1895)
Burns, A. R., *Warring States of Greece* (N.D.)
Church, A. J., *Helmet and Spear* (1900)
Cox, Sir G., *The Athenian Empire* (1876)
Griffiths, C. T., *The Mercenaries of the Hellenistic World* (Oxford, 1933)
Henderson, B. W., *The Great War Between Athens and Sparta* (1927)
Hignet, C., *Xerxes' Invasion of Greece* (1923)
Homer, *The Iliad* (1950)
Marsden, F. W., *Greek and Roman Artillery* (1969)
McKenzie, C., *Marathon and Salamis* (1934)
Parke, H. W., *Greek Mercenary Soldiers* (1933)
—— *Greek Mercenary Soldiers Fron Earliest Times to the Battle of Ipsus* (1933)
Plutarch, *Nine Greek Lives*
Snodgrass, A. M., *Arms and Armour of the Greeks* (1967)
Tarn, T. W., *Hellenistic Military and Naval Developments* (1950)

Thucydides, *Peloponnesian War*
Xenophon, *Cypruadia (Organisation and Tactics of the Persians and Greeks)*
—— *Anabasis*
—— *Cyropaedia*

THE SPARTANS

Forrest, W. G. G., *A History of Sparta 950–192 B.C.* (1968)
Sankey, C., *The Spartan and Theban Supremacies* (1877)

THE MACEDONIANS

Adcock, F. E., *The Greek and Macedonian Art of War* (1957)
Arrianus, Flavius, *The Life of Alexander the Great*
Dance, E. H., *Alexander the Great* (1948)
Dodge, T. A., *Alexander: History of the Origin and Growth of the Art of War* (1890)
Fuller, J. F. C., *The Generalship of Alexander the Great* (1958)
Hamilton, J. R., *Alexander the Great* (196?)
Lamb, Harold, *Alexander of Macedon* (1955)
Milns, R. D., *Alexander the Great* (1968)
Robinson, C. A., *Alexander the Great* (1963)
Tarn, W. W., *Alexander the Great* (1948)
Wheeler, B., *Alexander the Great* (1900)

THE ROMANS

Adcock, F. E., *The Roman Art of War under the Republic* (1940)
Appian, *Civil Wars*
Barker, H. (Baines and Noble), *Roman Legions* (1960)
Barrow, R. H., *The Romans* (1949)
Bitblade, L. O., *The Roman Invasions* (1935)
Buchan, J., *Julius Caesar* (1932)
Caesar, Julius, *Commentaries*, trans. Warrington, J. (1953)
Clark, J., *The Military Institution of the Romans* (1960)
Dodge, T. A., *Caesar: A History of the Art of War in Rome* (1892)
Forestier, Amadea, *Soldiers of Rome* (1928)
Fowler, W. W., *Julius Caesar* (1892)
Fuller, Major-Gen. J. F. C., *Julius Caesar—Man, Soldier, and Tyrant* (1965)
Gibbon, E., *Decline and Fall of the Roman Empire* (1872)
Grant, M., *Julius Caesar* (October, 1969)
Graves, Robert, *I Claudius* (Penguin, 1941)
Handford, S. A. (translated by), *The Conquest of Gaul* (Penguin, 1951)
Jones, A. H. M., *The Later Roman Empire, 284–602* (A.D.)
Kloss, F., *The Legion Marches* (N.D.)

Liddell-Hart, B. H., *Greater Than Napoleon, Scipio Africanus* (1926)
Livy, *Early History of Rome*
Marsden, E. W., *Greek and Roman Artillery* (1969)
Mellerch, H. E. L., *Soldiers of Rome* (1964)
Parker, H. M. D., *The Roman Legions* (1958)
Pelham, H. F., *Outlines of Roman History* (1922)
Plutarch, *Six Lives*
—— *Fall of the Roman Republic*
Polybius, *Histories*
Pratt, F., *Hail, Caesar* (1938)
Procopius, *History of the Wars*
Sallust, *The Legions*
Scullard, H. H., *Scipio Africanus, Soldier and Statesman* (1950)
—— *The Second Punic War* (1930)
Smith, R. B., *Rome and Carthage* (1887)
Tarn, Sir W., and Charlesworth, M. P., *Octavius, Anthony and Cleopatra* (1946)
Treece, Henry, *Legions of the Eagle* (1954)
Vegetius, *The Military Institutions of the Romans* (1960)
Walbank, F. W., *The Awful Revolution; The Decline of the Roman Empire in the West* (1968)
Warner, Rex, *Imperial Caesar* (1948)
Watson, G. R., *The Roman Soldier* (1969)
Webster, Graham, *The Roman Imperial Army* (1969)

### THE CARTHAGINIANS

Baher, G. P., *Hannibal* (1929)
Cottrell, Leonard, *Enemy of Rome* (1962)
De Beer, Sir G., *Hannibal* (1969)
Dodge, T. A., *Hannibal* (1891)
Haight, E. H., *Carthage and Hannibal* (N.D.)
Lamb, H., *Hannibal* (Bantam Book, 1958)
Livy, *The War With Hannibal*
Picard, G., and Charles, C., *Daily Life at Carthage in the Time of Hannibal* (N.D.)
Smith, R. B., *Carthage and the Carthaginians* (1916)
Torr, Cecil, *Hannibal Crosses the Alps* (1924)
Warmington, B. H., *Carthage* (Pelican A598, 1964)
Wilkinson, S., *Hannibal's March* (1911)

### THE BRITONS

Belloc, H., *Warfare in England* (1913)
Broxholme, J. F., *The War Queen* (1968).
Collingwood, R. G., *Roman Britain* (1923)
Cottrell, L., *The Great Invasion* (1958)
—— *Seeing Roman Britain* (1967)

Divine, D., *The North West Frontier of Rome* (*A Military Study of Hadrian's Wall*) (1969)
Dudley, D. R., and Webster, G., *The Roman Conquest of Britain, A.D. 43–57* (1965)
—— *Rebellion of Boudicca* (1962)
Durant, G. N., *Britain—Rome's Most Northerly Province, A.D. 43–450* (1969)
Hawkes, J., *Early Britain* (1946)
Jones, W. P., *The Conquest and Settlement of the Island of Britain* (1872)
Lindsay, J., *The Romans Were Here. The Roman Period in Britain and Its Place in our History* (1969)
Richmond, L., *Roman Britain* (1947)
Tacitus, *On Britain and Germany*
Vine, Rev. F. T., *Caesar in Kent* (1886)
Wilcox, H. A., *The Woodlands and Marshlands of England* (1933)
Windle, B. C. A., *The Romans in Britain* (1923)

### THE SAXONS

Abbott, J., *History of King Alfred the Great* (1898)
Baker, G. P., *The Fighting Kings of Wessex* (1931)
Clemoe, P., *The Anglo-Saxons* (1952)
Helm, P. J., *Alfred the Great* (1963)
Hollister, C. W., *Alfred the Great* (1965)
—— *Anglo-Saxon Military Institutions* (1968)
Major, A. F., *The Early Wars of Wessex* (1913)
Oman, C., *England Before the Norman Conquest* (1949)
Porter, H. M., *The Saxon Conquest of Somerset and Devon* (1967)
Stenton, Sir F. M., *Anglo-Saxon England* (1943)
Wheeler, B., *Alfred the Great* (1900)

### THE VIKINGS

Bronsted, Johannes, *The Vikings* (1960)
French, A., *Grettir The Strong* (1961)
Jones, Gwyn, *Nine Icelandic Sagas* (1961)
Magnusson and Palssan, *Njals Saga* (1960)
Sturluson, Snorri, *Heimsknngla—Norse King Sagas* (1961)

### THE BYZANTINES

Baynes, N., *The Byzantine Empire* (1925)
Graves, Robert, *Belisarius* (1938)
Hussey, J. M., *The Byzantine World* (1967)
Runciman, S., *Emperor Romanus Lecapenus and His Reign* (1929)

### THE GOTHS

Hodgkin, T., *Theodoric the Goth* (1890)

### THE FRANKS

Dymond, D., *Charlemagne* (N.D.)
Lamb, H., *Charlemagne* (1954)

### THE NORMANS

Adam, R., *A Conquest of England* (1965)
Barclay, Brig. C. N., *Battle 1066* (1966)
Barlow, Frank, *The Feudal Kingdom of England, 1042–1216* (1955)
Beeler, J., *Warfare in England, 1066–1189* (1966)
Belloc, H., *Warfare in England* (1913)
Brown, R. A., *The Normans and the Norman Conquest* (1969)
Butler, D., *1066 The Story of a Year* (1966).
Curtis, E., *Roger of Sicily and the Normans in Lower Italy, 1016–1154* (1912)
Davis, H. W. C., *England Under the Normans and Angevins, 1066–1279* (1909)
Douglas, D. C., *William the Conqueror* (1964)
Furneaux, R., *Conquest 1066* (1966)
Haskins, C. H., *The Normans in European History* (1916)
Heyer, G., *The Conqueror* (1931)
Hodgkin, R. H., *A History of the Anglo-Saxons* (1953)
Hollister, C. W., *Anglo-Saxon Military Institutions* (1962)
—— *The Military Organization of Norman England* (1968)
Loyn, Henry, *The Norman Conquest* (1965)
Muntz, H., *The Golden Warrior* (1962)
Penn, D., *Norman Castles in Britain* (1968)
Slocombe, G., *William the Conqueror* (1959)
Whiteoak, Douglas, Lemmon and Barlow, *The Norman Conquest* (1966)
Teach Yourself History Series, *William I and the Norman Conquest* (1966)

### THE CRUSADERS AND THE ARABS

Archer, T. A., and Kingsford, C. L., *The Crusades* (1894)
Archer, T. R., *Richard the Lion Hearted* (1888)
Barber, Richard, *The Knight and Chivalry* (1970)
Barker, Sir E., *The Crusades* (1923)
Belloc, H., *The Crusades* (1937)
Bradford, D., *The Great Betrayal: Constantinople 1204*
Brookman, Eric, *The Two Sieges of Rhodes* (1969)
Duggan, A., *The Story of the Crusaders* (1963)
Ellert, G., *The Knights of St John* (1958)

Engel, C. E., *Knights of Malta* (1963)
Fedden, R., *Crusader Castles—A Brief Study of the Military Architecture of the Crusade* (1950)
Fedden, R., and Thomson, J., *Crusader Castles* (1947)
Gabrieli, F., *Arab Historians of the Crusades* (1969)
Glubb, J. B., *The Great Arab Conquests* (1963)
Lamb, H., *The Crusades* (1931)
—— *Suleiman the Magnificent* (1953)
Lane-Poole, S., *Saladin* (1898)
Lawrence, T. E., *Crusader Castles* (1936)
Pernoud, R., *In the Steps of the Crusaders* (1959)
Perry, C. G., *History of the Crusades* (189?)
Rosebault, Charles J., *Saladin, Prince of Chivalry* (1930)
Runciman, Steven, *The History of the Crusades* (1951–4)
—— *The Kingdom of Acre* (1954)
—— *Fall of Constantinople* (1955)
Smail, R. C., *Crusading Warfare (1097–1193)* (1956)
Treece, Henry, *The Crusades* (1962).

## EASTERN WARFARE

*Mongols*

Grousset, René, *The Conqueror of the World, Chingis Khan* (1966)
Hookman, H., *Tamurlaine The Conqueror* (1962)
Lamb, H., *Babur The Tiger—1st Of The Great Moguls?* (1962)
—— *Genghis Khan, Emperor of all Men* (1928)
—— *Tamerlane* (1929)
—— *The March of the Barbarians* (1941)
Liddell-Hart, H., *Mongol Campaigns* (1929)
Phillips, E. D., *The Mongols* (1969)
Prawdin, M., *The Mongul Empire* (1952)
Van, V. (trans. Britton, L. E.), *Batu Khan* (1945)
—— (trans. Britton, L. E.), *Jenghis Khan* (1943)
Waley, A., *The Secret History of the Mongols* (1964)

*Huns*

Brien, M., *Attila, The Scourge of God* (1929)

*Japan*

James, D. H., *The Rise and Fall of the Japanese Empire* (1951)
Norman, F. J., *The Fighting Man of Japan-Samurai* (1905)
Varley, H. Paul (with Ivan and Nobuko Morris) *The Samurai* (1970)

*India*

Dikshitar, V. R. R., *War in Ancient India* (1948)
Irvine, W., *The Army of the Indian Moghuls* (1903)
Piggott, S., *Prehistoric India* (1952)

Sarkar, Jadanath, *Military History of India* (1960)
Sen, Surendranath, *The Military System of the Marathas* (1928)
Toy, S., *The Fortified Cities of India* (1965)
Watson, A., *The War of the Goldsmith's Daughter; The Moslem Conquest of Southern India* (1964)


### HUNDRED YEARS WAR AND MEDIEVAL PERIOD


Ashdown, C. G., *British Castles* (1911)
—— *Armour and Weapons in the Middle Ages* (1925)
Barber, Richard, *The Knight and Chivalry* (Longman, 197?)
Barrett, C. R. B., *Battles and Battlefields in England* (1896)
Beamish, T., *Battle Royal—Sime de Montfort's Struggle Against King Henry III* (1965)
Belloc, H., *British Battles—Crécy* (1912)
—— *British Battles—Poitiers* (1913)
Brooke, J. Little, *Knights of the Middle Ages* (1966)
Bryant, A., *Age of Chivalry* (1963)
Burne, A. H., *Crécy War* (Eyre and Spottiswood, 1965)
Burne, Lt. Col. A., *The Agincourt War. A Military History of the Latter Part of the Hundred Years War* (1956)
Clephan, R. C., *The Defensive Armour and the Weapons and Engines of War of Medieval Times and the Renaissance* (1900)
Conyn, M., *The Black Prince, 1330–1376* (1934)
Costain, R., *The Last Plantagenets* (1962)
Doyle, A. Conan, *The White Company* (1905)
—— *Sir Nigel* (1906)
Evans, Herbert A., *Castles of England and Wales* (1912)
Ffoulkes, Charles, *Armour and Weapons* (1909)
Fowler, Ken, *The King's Lieutenant—Henry of Grosmont, First Duke of Lancaster, 1310–61* (1969)
Hewitt, H. J., *The Black Prince's Expedition 1355–7* (1958)
Henty, G. A., *In Freedom's Cause* (1890)
—— *At Agincourt* (1897)
—— *Both Sides of the Border* (1899)
—— *St George for England* (1896)
Hibbert, Christopher, *Agincourt* (1964)
Hogg, G., *Castles of England* (1970)
Hogg, Brig. O. F. G., *English Artillery—1326–1716* (1963)
Keen, M. H., *The Laws of War in the Late Middle Ages* (1965)
Lindsay, Philip, *Henry V* (1934)
Mackenzie, Wm., *The Battle of Bannockburn* (1913)
McKisack, M., *Fourteenth Century* (Oxford Press, 1959)
Marzials, F. T., *Passages from Froissart* (1900)
Morris, J. E., *Welsh Wars of Edward I* (1897)
Nicholas, Sir H., *History of the Battle of Agincourt* (1971)
Oakshott, R. E., *The Sword and the Age of Chivalry* (1964)

Oakshott, R. E., *A Knight and His Armour* (1965)
—— *A Knight and His Horse* (1965)
—— *A Knight and His Weapons* (1965)
Oman, C., *The Art of War in the Middle Ages, 378–1515* (1885)
Oman, Sir C., *The History of the Art of War in the XIV Century* (1924)
O'Neil, B. H. S. and J., *Castles and Cannons* (1960)
Pearson, H., *Henry of Navarre* (1963)
Perroy, Edward, *The Hundred Years War* (1951)
Powicke, Sir M., *Battle of Lewes* (1953)
—— *Military Obligation in Medieval England* (1952)
Sellman, *Mediaeval English Warfare* (1960)
Simpson, Douglas, *Castles of England and Wales* (1969)
Teach Yourself History Series, *Henry V and the Invasion of France* (1947)
Thompson, A. H., *Military Architecture in England During the Middle Ages* (1912)
Toy, Sidney, *The Castles of Great Britain* (1953)
—— *Castles: A Short History of Fortifications from 1600 B.C. to A.D. 1600* (1959)
Warner, Philip, *Sieges of the Middle Ages* (1968)

## APPENDIX 4

# Relevant Articles in Magazines and Journals

Material, often slanted in the direction of wargaming, can be found in articles written by military historians and wargamers and published in journals directly with wargaming or with a 'fringe' connection. The list that follows details articles that have appeared in such publications during the last fifteen years and no doubt those magazines still in production will continue to publish material not only of value to the wargamer specialising in the Ancient period but also concerned with the hobby in general. Some of these journals are no longer in production and their publishers will long since have disposed of all available copies. Nevertheless, it is not beyond the scope of the enthusiastic wargamer to discover a fellow hobbyist whose bookshelves hold the treasured volumes.

The availability of these journals mentioned is as follows:

*The Armchair General*, P.O. Box 274, Beltsville, Maryland 20705, U.S.A. (may be out of production)
*Bayonet*, Hamish Fraser, 27 Ramsgate Road, Margate, Kent (may be out of production)
*Bristol Wargames Society Journal*, Steve Curtis, 10 Hurle Crescent, Clifton, Bristol BS8 2TA
*The Bulletin of the British Model Soldier Society* (bi-monthly to members), B. A. Hornick, 16 Charlton Road, Kenton, Middlesex
*The Courier*, R. Bryant, 45 Willow Street, Brockton, Mass., U.S.A.
*The Gazette of the Ox/Bucks and Berks Branch British Model Soldier Society*, 33 Wentworth Road, Oxford.
*History Today,* Bracken House, 10 Cannon Street, London E.C.4
*Milihistriot*, Bob Bard, Box 1463, Baltimore, Maryland 21203, U.S.A.
*Miniature Parade*, U.S.A. (out of production)
*Miniature Warfare*, 1 Burnley Road, Stockwell, London S.W.9
*Slingshot—Official Journal of the Society of Ancients*, Philip Barker, 757 Pershore Road, Selly Park, Birmingham 29
*Soldier*, 433 Holloway Road, London N.7
*Strategy and Tactics*, Box 396, New York 10009, U.S.A.
*Table Top Talk*, U.S.A. (out of production)
*Tradition*, 188 Piccadilly, London W.1

*The Vedette—Journal of the National Capital Military Collectors*, P.O. Box 30003, Bethesda, Maryland 20014, U.S.A.

*Wargamer's Newsletter*, D. Featherstone, 69 Hill Lane, Southampton, SO1 5AD

*Wargames Digest*, U.S.A. (out of production)

'Ancient Warfare' by T. Bath, *Wargamer's Newsletter*, February 1964

'Ancient Warfare' by P. Barker, *Miniature Warfare*, November 1968

'Ancient Warfare' by R. Turner, *Bristol Wargames Society Journal*, Spring 1970

'Horses, Arrows and Ancient War' by A. Nickels, *Slingshot*, September 1967.

'Infantry Tactics in the Ancient World', *Wargamer's Newsletter*, June 1967

'Mediaeval Troops' by T. Bath, *Wargamer's Newsletter*, December 1965

'Mediaeval Warfare' by N. Chamberlin, *Table Top Talk*, March 1965

'Wargames in the Holy Land' by C. A. Sapherson, *Wargamer's Newsletter*, May 1967

'Ancient Armies in Miniature' by D. Rusk, *The Miniature Parade*, Winter 1967

'Ancient Period Armies' by H. Gerry, *Wargamer's Newsletter*, April 1965

'Ancient Wargames Armies and Tactics' by P. Barker, S. Reed and T. Dixon, *Slingshot*, July 1970

'A Cheap Way to Collect Ancients and Mediaevals' by T. Bath, *Wargamer's Newsletter*, May 1965

'An Ideal Ancient Army' by H. Gerry, *Slingshot*, May 1966

'Organization of Ancient Wargames Armies' by T. Bath, *War Games Digest*, December 1957

'Organisation of a Mediaeval Wargames Army' by O. Sorosoto, *Table Top Talk*, May 1965

'Romans v Gauls in Wargames' by J. Scruby, *Table Top Talk*, February 1962

'Ancient Wargaming' by P. Barker, *Miniature Warfare*, November 1968

'Ancient Wargaming' by E. P. Smith, *Miniature Warfare*, May 1968

'Generalship in Ancient Wargames' by T. Bath, *War Games Digest* December 1957

'An Introduction to Ancient Wargaming' by P. Barker, *Miniature Warfare*, March 1968

'The Medieval Battle Game' by J. E. Dotson, *Strategy and Tactics*, January–February 1969

'Medieval Wargames' by C. Sweet, *War Games Digest*, December 1957

'Thoughts on Ancient Wargaming' by D. Millward, *Wargamer's Newsletter*, November 1969

'Wargames in the Ancient Period' by T. Bath, *War Games Digest*, December 1957

'Ancient Rome—Rules Part I and II' by J. Greer, *Wargamer's Newsletter*, May/June 1969

'Ancient Wargames Rules by Archimedes', *War Games Digest*, March 1961, September 1961 and March 1962
'Ancient Wargame Rules' by P. Barker, *Wargamer's Newsletter*, June 1966
'The Dead, Dying and Injured' (estimating Wargaming Casualties in Medieval Period) by K. Bastian, *Wargamer's Newsletter*, April 1964
'Mediaeval Campaign Rules' by T. Bath, *Wargamer's Newsletter*, December 1963
'My Ancient Combat Method' by C. A. Cass, *Wargamer's Newsletter*, October 1969
'New Ancient Campaign Rules' by T. Bath, *Wargamer's Newsletter*, October 1963

'Ancient Naval Warfare' by T. Bath, *Wargamer's Newsletter*, April 1966

'Fighting Wargames in the Hyborean Age' by A. W. Saunders, *War Games Digest*, December 1960

'Ancient Missile Weapons' by B. O'Brien, *Wargamer's Newsletter*, August 1968
'Artillery of the Ancients' by T. Bath, *Slingshot*, July 1966
'Bows and Arrows', *Soldier*, February 1967
'Cavalry Weapons' by P. Barker, *Slingshot*, September 1966
'Characteristics of Ancient Weapons—Part I—Infantry' by P. Barker, *Slingshot*, September 1966.
'Comparison of English Longbows and Turkish Composite Bows' by P. Barker, *Miniature Warfare*, September 1968
'Crossbows and Longbows' by P. McManus, *Slingshot*, January 1968
'Heavy Infantry Spears' by P. Barker, *Slingshot*, May 1968
'Longbows v Horse Archers' by W. T. Thurbon, *Wargamer's Newsletter*, December 1967
'Mail Armour' by P. Woolacott, *Slingshot*, July 1970
'Pole Arms' (author not named), *Bulletin*, no. 10, December 1958
'Some Notes on Crossbows and Longbows' by E. P. Smith, *Miniature Warfare*, March 1968 and April 1968
'Some Reflections on Ancient Missile Weapons' by R. B. Nelson, *Wargamer's Newsletter*, June 1968
'Thoughts on Archers in Wargames' by R. O'Brien, *Slingshot*, May 1967

'Camels at War' by D. Rusk, *Table Top Talk*, July 1965
'Cataphract Cavalry' by P. Barker, *Slingshot*, July 1967
'The Horse in Ancient Warfare' by A. L. Nickels), *Slingshot*, July 1969, September 1969 and November 1969
'Wagon Laagers' by Antistentor, *Slingshot*, January 1971

'Battle for Maiden Castle' by M. Bath, *War Games Digest*, March 1961
'Slingers at Maiden Castle' by A. Colquhoun, *Table Top Talk*, April–May 1963

'The Ancient Egyptian Chariot' by A. Buttery, *Slingshot*, May 1971
'Chariots' by C. Sweet, *War Games Digest*, December 1961

'Egyptian War Chariot' by D. Guyler, *Bulletin*, January, February, March 1960
'Egyptian War Chariot' by R. B. Nelson, *Slingshot*, January 1971
'War Chariots of the Egyptians' by D. Guyler, *Slingshot*, May 1966

'Alexander the Great's Iron Elephants', *Wargamer's Newsletter*, December 1967
'Armour for an Elephant' (picture), *Soldier*, November 1961
'Concerning Elephants'—*Armchair General:* Vol. I. No. 6
'Elephants' by P. Condray—*Wargamer's Newsletter,* May 1966
'Elephants' by R. King, B. Long and D. Taylor, *Slingshot*, January 1971
'Elephants in Wargames' by A. C. Cass, *Wargamer's Newsletter*, September 1963
'The Use of Elephants in the Ancient World' by A. C. Rogers, *Slingshot*, March 1967
'Wargaming with Elephants' by H. Gerry and T. Bath, *Slingshot*, September 1966

'Greek Siegecraft Before 400 B.C.' by R. B. Nelson, *Slingshot*, July 1970
'Medieval Sieges as a Wargame' by J. L. Mumford, *Slingshot*, July 1966
'Siege Engines' by T. Bath, *Wargamer's Newsletter*, February 1966
'Siege Warfare' by T. Bath, *Wargamer's Newsletter*, February 1963

'Ancient Armies in Miniature (Ancient Egypt)' by D. Rusk, *The Miniature Parade*, Winter 1967
'Ancient Armies in Miniature (Assyria)' by D. Rusk, *The Miniature Parade*, Winter 1967
'The Ancient Assyrians' by Cassius Maximus, *British Model Soldier Society Bulletin*, January 1959
'The Ancient Egyptian Army' by N. Dickinson, *Slingshot*, July 1968
'Ancient Warriors' (illustrations) by J. Blanche, *Slingshot*, May 1971
'Assyria—The Warrior Nation' by D. Kromm, *Slingshot*, November 1969
'The Assyrian Army' by N. Dickinson, *Slingshot*, November 1968
'Assyrian War Chariots' by D. Guyler, *Slingshot*, November 1965
'Assyrian Warriors and War Engines' (illustrations), *Slingshot*, January 1971
'The Assyrians' by A. C. Cass, *Slingshot*, November 1970
'The Assyrians' by B. Reilly, *Slingshot*, May 1967
'Battle of Kadesh 1293 B.C. (Egyptians v Hittites)' by M. Bath, *War Games Digest*, September 1961
'Biblical Warfare—The Invasion of Canaan' by D. Millward, *Slingshot*, July 1969, September 1969, November 1969 and January 1970
'Persian and Assyrian Costume' by J. Blanche, *Slingshot*, January 1971

'The Army of Sassanid Persia' by J. Haldon, *Slingshot*, May 1968
'Ancient Armies in Miniature (The Persian Empire)' by D. Rusk, *The Miniature Parade*, Winter 1967

'Ancients (Persians 480 B.C. Uniforms)' by P. Abbott, *Bulletin*, May 1962
'Battle of Arbela as a Wargame' by H. Gerry, *Slingshot*, July 1969
'The Campaign and Battle of Salamis' by T. Bath, *Slingshot*, March 1966
'Greek and Persian Costume' by R. Webb, *Slingshot*, March 1966
'Medes and Persians' by D. Oakley, *The Gazette* (Journal Ox/Bucks Branch British Model Soldier Society), August 1970
'A Note on the Army of Xerxes' by A. C. Rogers, *Slingshot*, September 1967
'The Persian Army' by N. Dickinson, *Slingshot*, November 1969
'The Persian Army 480 B.C.' by J. M. Steel, *Slingshot*, September 1970
'Persian Wargames Armies' by A. Cass, *Slingshot*, November 1970
'Seleucid Armies' by D. Millward, *Slingshot*, January 1971
'Some Seleucid Wargames Armies' by R. Goodwell Boss, *Slingshot*, September 1970

'Ancient Armies in Miniature (The Greeks)' by D. Rusk, *The Miniature Parade*, Winter 1967
'The Battle of Arbela as a Wargame' by H. Gerry, *Slingshot*, July 1969
'The Athenian Navy' by R. B. Nelson, *Slingshot*, September 1968
'The Campaign and Battle of Salamis' by T. Bath, *Slingshot*, March 1966
'Greek Armour' by R. Webb, *Slingshot*, March 1970
'Greek Hoplites' by R. B. Nelson, *Slingshot*, March 1971
'Greek Officers and Standards' by R. Webb, *Slingshot*, November 1970
'Greek and Persian Costume' by R. Webb, *Slingshot*, March 1966
'Greek Shields' by R. Nelson, *Slingshot*, November 1970
'Greek Siegecraft before 400 B.C.' by R. B. Nelson, *Slingshot*, July 1970
'Greek Wargames Armies' by A. C. Cass, *Slingshot*, November 1970
'The Hoplite' (illustrations) by P. Wilcox, *The British Model Soldier Society Bulletin*, No. 3, June 1970
'Hoplite Shield Blazons' by R. B. Nelson, *Slingshot*, November 1970
'Hoplite Shields and Spears' by A. C. Nickels, *Slingshot*, March 1969
'Battle of Leuctra 371 B.C.', *Wargamer's Newsletter*, April 1969
'Notes and Queries from Zenophon' by C. Grant, *Slingshot*, May 1970
'Notes on Xenophon' by R. Nelson, *Slingshot*, September 1970
'Shield Emblems for Hoplites' by C. M. Dorman, *Slingshot*, July 1970
'Thracian Warfare' by Antistentor, *Slingshot*, March 1971

'Phalanxes in Ancient Warfare' by H. Gerry, *Miniature Warfare*, June 1970
'The Spartan Army' by R. B. Nelson, *Slingshot*, July 1968
'A Spartan Hoplite' (illustration) by R. B. Nelson, *The British Model Soldier Society Bulletin*, August 1969
'The Spartans' by P. Abbott, *The British Model Soldier Society Bulletin*, February 1963

'Alexander The Great' by E. Badian, *History Today*, June 1958
'Ancient Armies in Miniature—The Macedonians' by D. Rusk, *The Miniature Parade*, Winter 1967
'The Battle of Hydaspes as a Wargame' by C. S. Grant, *Slingshot*, September 1970
'The Macedonian Phalanx' by J. M. Hinds, *The Armchair General*—vol. I, no. 5; vol. II, no. 2; vol. II, no. 3
'The Phalanx' by A. C. Cass, *Slingshot*, March 1971
'Phalanx v Phalanx' by R. B. Nelson, *Wargamer's Newsletter*, August 1967

'The Roman Army 753 B.C.–1453 A.D.' by S. Patrick, *Strategy and Tactics*, January–February 1971
'A Concise History of the Known Legions' by A. C. Cass, *Slingshot*, July 1966, September 1966, November 1966
'The Roman Armies' by Cassius Maximus, *The British Model Soldier Society Bulletin*, May 1957
'The Roman Army' by L. G. Morgan, *Miniature Warfare*, December 1968
'Roman Army Organisation' by H. Pestana, *Table Top Talk*, March 1965
'The Composition and Tactics of the Roman Army' by R. Rotherham, *Miniature Warfare*, December 1969
'Some Notes on the Roman Army' by G. Sandford, *War Games Digest*, Winter 1962
'The Imperial Roman Army and its Enemies', *Miniature Warfare*, May 1969
'The Imperial Roman Legions' by M. Simkins, *Tradition*, nos. 41, 42 43 and 46
'The Roman Legions' by Cassius Maximus, *The British Model Soldier Society Bulletin*, March 1957
'The Roman Legion' by Cassius Maximus, *The British Model Soldier Society Bulletin*, February–April 1970
'The Roman Legion' by D. Rusk, *War Games Digest*, Winter 1962
'Under Roman Eagles' by J. R. Hinds, *Armchair General*, vol. I, no. I
'The Strength of Roman Units After Hadrian' by P. Barker, *Slingshot*, September 1970
'The Roman Army of the 4th and 5th Centuries' by P. Barker, *Slingshot*, November 1968 and January 1969
'The Later Roman Army' by J. Hinds, *Vedette*, February 1968, June 1968 and August 1968
'Roman Field Tactics—Part 1 and 2' by J. Hinds, *Armchair General*—vol. I, no. 3, January 1969; vol. I, no. 4, February 1969
'Roman Cavalry' by Antistentor, *Slingshot*, January 1971
'Roman Soldiers' by B. Bard, *Milihistriot*, October 1965
'Arms and Equipment of the Roman Soldier' by L. M. Patt, *Courier*, vol. 2, no. 1

'Standards of the Roman Legions' by J. R. Hinds, *Armchair General*, vol. 1, no. 2
'The Roman Pilum and Hasta' by A. C. Cass, *Slingshot*, March 1966
'Roman Military Awards' by D. Guyler, *Slingshot*, November 1966
'Ranks in the Roman Army' by Cassius Maximus, *The British Model Soldier Society Bulletin*, April, June, August and October 1969
'Figures on Trajan's Column' by R. Johnson, *Slingshot*, March 1971
'The Roman Catapult' by P. Loughran, *Bayonet*, vol. 2, no. 1
'A Few Comments (Roman Uniforms)' by Cassius Maximus, *The British Model Soldier Society Bulletin*, November 1957
'Roman Uniforms' by G. Sandford, *War Games Digest*, Summer 1962
'Roman Clothing in 3rd Century A.D.' by P. Barker, *Slingshot*, July 1970
'The Roman Army Medical Corps' by D. Guyler, *Slingshot*, November 1968
'The Roman Army Medical Corps' by D. Guyler, *The British Model Soldier Society Bulletin*, February 1961
'The Praetorian Guard' by B. Reilly, *Slingshot*, March 1971
'The Praetorian Guard' by G. Powell, *History Today*, December 1968
'Cannae 216 B.C.' by R. Johnson, *Slingshot*, March 1971
'The Campaign and Battle of Cannae' by M. Kleiderman, *Slingshot*, January 1966
'Campaign and Battle of Carrhae 53 B.C.' by T. Bath, *Slingshot*, May 1966
'Battle of Pharsalia 48 B.C.' (Julius Caesar and Pompey) by C. S. Grant, *Slingshot*, March 1970
'The Threat to Rome 120–101 B.C.' by J. L. Railton, *Slingshot*, May 1970
'The Defeat of Cestius Gallus A.D. 68' by S. G. F. Brandon, *History Today*, January 1970
'Arrian's Orders of March and Battle against the Alains' by P. Barker, *Slingshot*, May 1969
'Roman Conquest of Spain' by Antistentor, *Slingshot*, January 1971
'Wargaming with a Roman Army' by S. Colby, *Slingshot*, January 1971
'Romans v Gauls as a Wargame' by J. Scruby, *Table Top Talk*, February 1962
'Imperial Caesar' by P. Green, *History Today*, October 1960
'Hadrian's Wall', *Soldier*, July 1969
'Spartacus' by J. S. Richardson, *The Gazette* (Journal Ox/Bucks Branch British Model Soldier Society), August 1970

'The Army of Carthage' (illustrated), *War Games Digest*, Winter 1962
'Notes on the Carthaginian Army' by R. O'Sullivan, *Slingshot*, January 1967
'Carthaginian Troops' by N. Driscoll, *Table Top Talk*, January 1965
'Carthaginian Wargames Army' by A. C. Cass, *Slingshot*, November 1970
'The Campaign and Battle of Cannae' by M. Kleiderman, *Slingshot*, January 1966

'Syracuse 415–413 B.C.' by T. Bath, *Slingshot*, July 1966
'Battle of Lake Trasimene' by R. Johnson, *Slingshot*, November 1970
'Battle of Zama as a Wargame' by C. Grant, *Slingshot*, May 1971

'Romans v Gauls in Wargames' by J. Scruby, *Table Top Talk*, February 1962
'The Galatian Invasions' (Gauls) by D. Taylor, *Slingshot*, March 1971

'Ancient Germanic Tribesmen' by C. B. Drake, *Miniature Warfare*, July 1970 and August 1970

'Boadicea and a Briton Chief', Prologue to Ancient British History by B. Bard, *Milihistriot*, October 1965
'The Dark Ages' (Anglo-Saxon Military Notes) by J. Tofts White, *The British Model Soldier Society Bulletin*, February 1957

'Goths—Wargame Army' by B. O'Brien, *Slingshot*, September 1970
'Adrianople' by P. Barker, *Slingshot*, November 1968

'The Celts', *The British Model Soldier Society Bulletin*, June–July 1959

'The Jomsvikings' by R. Goodwell-Boss, *Slingshot*, September 1970

'The Mongols' by K. Sherlock, *Slingshot*, July 1966 and November 1966
'The Mongol Army' by D. Rusk, *Table Top Talk*, November 1963
'The Mongol Army' by D. MacFarlane, *Miniature Warfare*, April 1969
'The Mongols and Europe' by J. A. Boyle, *History Today*, May 1959

'The Normans' by R. Goodwell-Boss, *Slingshot*, November 1970
'1066 And All That' by T. Bath, *Wargamer's Newsletter*, June 1966
'Hastings Reconstructed' by D. Featherstone, *Wargamer's Newsletter*, March 1966
'Soldiers at Hastings' by R. C. Lawson, *Miniature Warfare*, February 1969
'The Norman Invasion of Ireland' by R. O'Sullivan, *Slingshot*, March 1967 and May 1967

'The Battle of Dorylaeum' by T. Bath, *Slingshot*, March 1968
'The Teutonic Knights' by A. L. Nickels, *Slingshot*, September 1966
'The Siege of Rhodes' by A. D. Lacey, *History Today*, May 1968
'The Great Siege of Malta' by T. H. McGuffie, *History Today*, August 1965
'The Turkish Janissaries' by Major W. Gordon Clark, *The British Model Soldier Society Bulletin*, October 1957

'The Byzantine Army' by H. J. Powell, *The Miniature Parade*, February 1969
'12th Century Byzantines' by C. Sweet, *War Games Digest*, June 1959
'Forming a Byzantine Wargames Army' by P. Barker, *Slingshot*, July 1967

'Edward I and Castles of North Wales' by A. Rogers, *History Today*, July 1969

'Development of Armour During 100 Years War' by L. Archer, *Tradition*, nos. 7, 8 and 10
'French Armour *c.* 1280', *Tradition*, no. 13
'French Crossbowmen of the 15th Century' by B. Bard, *Milihistriot*, January 1968
'Some English Coats of Arms of the Mid 14th Century' by A. L. Nickels, *Slingshot*, January 1968
'French Coats of Arms of Mid 14th Century' by A. L. Nickels, *Slingshot* March 1968
'Coats of Arms at Agincourt' by A. L. Nickels, *Slingshot*, July 1968 and September 1968
'The Campaign and Battle of Agincourt' by T. Bath, *Slingshot*, November 1965
'The Battle of Agincourt' by A. H. Burne, *History Today*, September 1956
'Churchill on Agincourt' by T. J. Deakin, *The Vedette*, October and December 1968
'John 'O Gaunts Grande Chevauchee 1373' by A. H. Burne, *History Today*, February 1959

'Swiss Confederates (1350–1500)' by K. Bastian, *War Games Digest* December 1957
'Warfare in the 16th Century—The Swiss' by P. F. Purton, *Miniature Warfare*, July 1970

'The Landsknecht' by J. Proctor, *Courier*, no. 2, vol. 2, 1966

'Irish Warfare 8th–14th Century' by R. King, *Slingshot*, January 1971

'Warfare in China' by K. Sherlock, *Slingshot*, March/May 1968
'Ancient Indians' by A. C. Cass, *Slingshot*, March 1971
'War In Vedic Period—Ancient India' by D. G. Smith, *Wargamer's Newsletter*, April 1966
'Ancient Japanese Wargames' by K. Sherlock, *Slingshot*, January 1966
'Oriental Warfare' by K. Sherlock, *Slingshot*, November 1965

'Mercenaries' by P. McManus, *Slingshot*, September 1967
'Mercenary Soldiers' by Col. H. C. B. Rogers, *Tradition*, no. 47/48
'Use of Mercenary Troops' by Russell Vallence, *Wargamer's Newsletter*, January 1970
'Mercenaries' by C. Wesencraft, *Wargamer's Newsletter*, January 1970
'The Varangian Guard' by G. Benson, *Slingshot*, January 1970

# APPENDIX 5
## General Notes on Weapons

At one or two points in the book a mention of specific weapons may be taken literally by readers and used in their wargaming forces. For example, 'Lance' is mentioned many times as being the equipment of cavalry of various peoples. The term should be reserved for a weapon such as the heavy Kontos used by the Sarmatians and similar. In fact, there were very few 'Lancers' as we understand them used by the Ancients, mainly because of the lack of stirrups, and more particularly, lack of a suitable saddle. The standard cavalry arm for a long time was a short spear plus in many cases a case of javelins, and of course, sword, light axe, or other close action weapon, although the short spear was probably still the most effective weapon for close action against enemy cavalry.

A 'Pike' is a two-handled weapon over 12 feet in length and used by troops in close order in controlled masses. Therefore it was not the weapon of barbarian 'wild men' and even less that of levies. Neither would be amenable to the type of training that would be needed to make a pike phalanx effective. Research Group Rules now classify the short and long spear types for infantry, and restrict spear armed bowmen to a short spear only.

# APPENDIX 6
# Wargames Research Group Ancient and Medieval Rules

Bob O'Brien, who has so ably 'vetted' this manuscript, has been a major figure in the compilation of the various sets of rules put out by the War Games Research Group, beginning with Ancient, through Medieval and lately into Horse and Musket and 20th century. It is with the first two sets of rules that we are concerned. Certain pertinent points arise from comments made by Mr O'Brien in relating this book and as they obviously present valuable information to Ancient and Medieval wargamers they are considered at some length in this appendix.

So far as Fighting Assessment is concerned, Mr O'Brien puts forward the following suggestions where Research Group Rules are being used.

(a) Types of troops who show deficiencies in any of the qualities of Order, Training and so on would be penalised in their Reaction to fresh circumstances, and their ability to understand and act on fresh orders. They should be made to respond more slowly, particularly to the Optional actions that are built into the rules.

(b) Those who are low on Bravery and Discipline would suffer on Reaction so that they would be less likely to stand a charge and even less likely to initiate one, and this would be accentuated if they were facing people of known worth and discipline or ferocity.

(c) There might well be consideration given to some kind of 'class distinction' between cavalry who, regular or barbarian, reckon themselves several degrees superior to most types of infantry. This not only applies to Medieval Times but is a steady thread running all through time when the horse was used in war. Fighting Assessment factors might well reflect this as between horse and foot in the same army, in varying degrees. Possibly the most important aspect where Fighting Assessment variations can be introduced is where they have an effect on Reaction because the tendency in these rules and their amendments is to emphasise the differences in fighting value and Reaction between differing types. Latest amendments to the rules make 'E' type troops have further disadvantages although 'costing less' when amassing as an army. 'A' 'B' types of troops (particularly infantry) although 'costing' more, have greater steadiness and it takes correspondingly longer to break them. Other factors in the rules tend to force a wargamer into using troops in the way they were meant to be used. For example, trying complicated manoeuvres with Barbarian types of troop under Research Group Rules is asking for disaster as Reaction tends to cause them to perform in a manner exactly opposite to what was intended!

War Games Research Group Rules place a great importance upon the classification of troops, insisting that no troops can be used in an army that were not included in its prototype. Also, they find it important to specify the proportions, for example an enthusiastic wargamer discovered that Alexander used light cavalry armed with the sarissa as a type of lancer, so he rushed off and formed units of them, not realising that only one unit of this type of troops was ever formed and it apparently did not distinguish itself because the experiment was not repeated. It is significant that the organisers of the 1971 War Games Club Championship Convention in Birmingham have been careful to specify in what classes and overall proportions different types of troops can be included.

In the Research Group Medieval Rules, the types of troops that can be used in various armies are prescribed and are classified according to capability and quality. Points values generally reflect the variations in types. For example, a Swiss Army operates in strictly traditional style, on the wargames table, push a French or German, both horse and foot, right off the wargames table. On the other hand, should the Swiss be unlucky enough to come against a Mongol or Turkish force in the open, then horse archers would whittle down the Swiss columns which would be unable to close with their faster moving opponents.

Mr O'Brien says that he would approve of differing value commanders under the Ancient rules (similar to the system of the Medieval) with corresponding effects on Reaction. 'Above-average' commanders would have a steadying effect, so making unreliable troops more controllable while the 'Below Average' commander would be either of the order of 'a gallant twit' who would enthuse troops (particularly cavalry) to useless charges, or 'poltroon' who would be generally not trusted and whose orders would bend to be disobeyed or only half performed, particularly by trained troops. Conversely, the loss of an 'Above Average' commander would be likely to have a greater effect on morale on most troops other than personal bodyguards, who would probably feel inclined to avenge his death. All these factors can be brought into the rules by judicious alterations to Reaction and other factors.

Further amendments to the War Games Research Group Rules will make allowances for Fanatics (including Berserks) in limited quantities with appropriate fighting and morale factors.

Finding the system of alternative orders in this manuscript to be 'complicated' Mr O'Brien states his prejudice in favour of orders given at the start and changed only by messenger, personal contact by command figure or signal. When the Research Group Ancient Rules were first produced a prime object was to avoid writing orders each game-move by the Orders and Reaction system. In an introduction by Philip Barker to the War Games Research Group Infantry Action Rules the question of writing orders each game-move is mentioned as '.... assuming each unit commander is a hero, clairvoyant and genius'.

# Index